D0420277

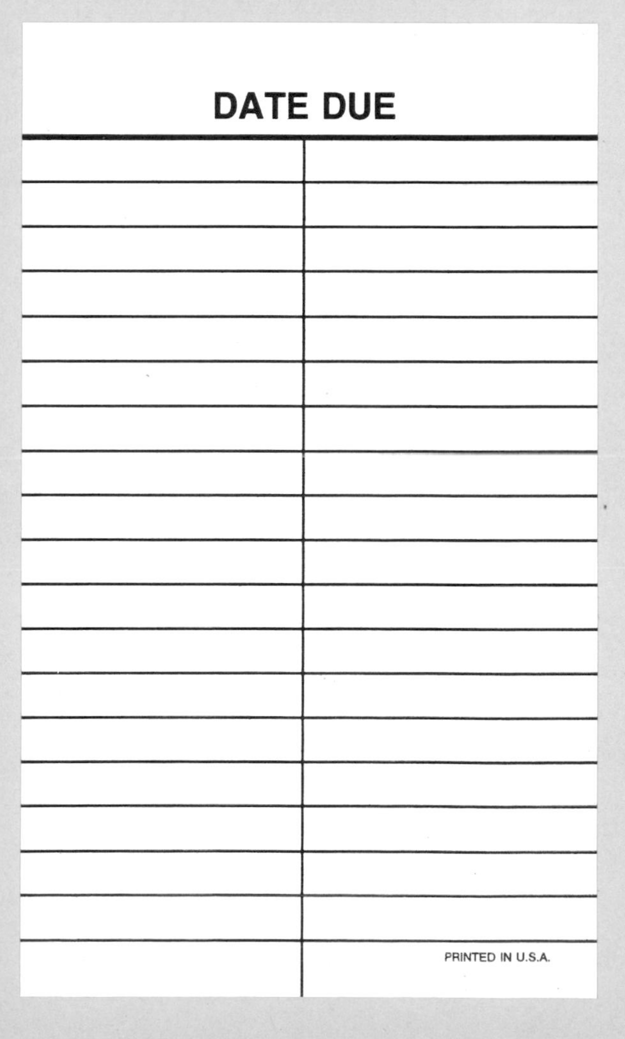
DATE DUE
PRINTED IN U.S.A.

The Rape of the Lock

and its illustrations

1714–1896

By the same author

(Co-Editor) *Essays and Poems and Simplicity, a Comedy*
by Lady Mary Wortley Montagu (1977)

(Editor) *Court Eclogs Written in the Year, 1716*
by Lady Mary Wortley Montagu
[Pope's autograph manuscript] (1977)

Lord Hervey: Eighteenth-Century Courtier (1973)

(Editor) *The Spectator* by Joseph Addison and Richard Steele
[a modern selection] (1970)

(Editor) *Selected Letters* by Lady Mary Wortley Montagu (1970)

(Editor) *Complete Letters* by Lady Mary Wortley Montagu
(three volumes, 1965–67)

The Life of Lady Mary Wortley Montagu (1956)

[William Hamilton], The Rape of the Lock [1790–1800], oil

The Rape of the Lock and its illustrations 1714–1896

ROBERT HALSBAND

CLARENDON PRESS · OXFORD

1980

Oxford University Press, Walton Street, Oxford OX2 6DP

OXFORD LONDON GLASGOW
NEW YORK TORONTO MELBOURNE WELLINGTON
KUALA LUMPUR SINGAPORE JAKARTA HONG KONG TOKYO
DELHI BOMBAY CALCUTTA MADRAS KARACHI
NAIROBI DAR ES SALAAM CAPE TOWN

Published in the United States
by Oxford University Press, New York

British Library Cataloguing in Publication Data

Halsband, Robert
'The rape of the lock' and its illustrations, 1714–1896
1. Pope, Alexander. Rape of the lock – Illustrations
I. Title
741.64 NC960 79-40481

ISBN 0-19-812098-2

Filmset and printed by BAS Printers Limited,
Over Wallop, Hampshire

PREFACE

It is an astonishing fact of literary history and criticism that Alexander Pope's *The Rape of the Lock*, one of the most famous and popular poems of the eighteenth century, has never been analysed in relation to the six full-page engraved plates with which it was illustrated on its first publication in 1714. In our own day, the most elaborate, scholarly treatment of the poem—the Twickenham Edition—simply records the names of the artist and the engraver, adding a single footnote: 'For these French artists see DNB [Dictionary of National Biography].'[1] Almost without exception, the many critical studies devoted to the poem in recent years, from Cleanth Brooks's brilliant pioneer essay to various 'casebooks', disregard its illustrations.[2] Two recent studies of Pope's interest in art do not even mention these illustrations.[3]

Yet, as Philip James has written, 'An illustrated book is a partnership between author and artist to which the artist contributes something which is a pictorial comment on the author's words or an interpretation of his meaning in another medium.'[4] This statement obviously applies to fiction, to such novels as Richardson's and Dickens's. How true is it, one wonders, for English poetry? Only Shakespeare, Milton, and James Thomson have been treated in

(Place of publication is London unless otherwise stated.)

1. *The Rape of the Lock and Other Poems*, ed. Geoffrey Tillotson, 3rd ed., The Twickenham Edition of the Poems of Alexander Pope, ed. John Butt, IV (1962), p. 140 n. And as an illustration it reproduces an engraving that is questionably ascribed to Hogarth and questionably dated 1717 (see p. 24 below).
2. One plate is very briefly mentioned by Earl R. Wasserman in 'The Limits of Allusion in *The Rape of the Lock*', *JEGP*, 65 (1966), 441, n. 41. More recently Clarence Tracy has made use of the plates in an edition of the poem 'illustrated by means of numerous pictures, from contemporary sources, of the people, places, and things mentioned' (*The Rape Observ'd*, Toronto, 1974). Besides disregarding the most provocative plate, the frontispiece, Professor Tracy does not investigate whether or how the illustrations go beyond the text— which is my main interest.
3. James Sambrook, 'Pope and the Visual Arts', *Writers and their Background: Alexander Pope*, ed. P. Dixon (1972); Morris R. Brownell, *Alexander Pope & the Arts of Georgian England* (Oxford, 1978).
4. *English Book Illustration 1800–1900* (1947), p. 7.

extended studies that trace the history of their illustrations.[5] (Blake, about whom commentary is superabundant, stands apart since he illustrated his own writings.) 'The critical history of a great literary work', a Miltonist has written, 'is incomplete unless it incorporates inferences from the interpretation put upon the masterpiece by artists.'[6] Certainly *The Rape of the Lock*, an acknowledged classic for more than 250 years, deserves such an examination.

Illustrations that appear with a literary work on its first publication are particularly significant, for they elucidate, modify, and supplement the meaning of the verbal text, particularly if the illustrations were drawn to the author's specifications or were subject to his approval. Those that Pope may have suggested (or accepted) for *The Rape of the Lock* supplement the poem in ways not explicitly set forth in the text. As for illustrations designed for the poem during a later period: they reflect contemporary critical attitudes toward Pope and his masterpiece; they also show how the taste and sensibility of that period, reflected through individual artists, responded to the text. *The Rape of the Lock* and its illustrations richly fulfill these possibilities, for not only was it illustrated on its initial publication in 1714, but it was elegantly 'adorned' at the end of the eighteenth, often illustrated during the nineteenth, and then 'embroidered' by a great artist on the eve of the twentieth century.

Having thus defined my approach, I had better add that it differs from one that in recent years has been taken in such books as Jean H. Hagstrum's *The Sister Arts* (1958) and Mario Praz's *Mnemosyne The Parallel Between Literature and the Visual Arts* (1970). Deriving from Horace's phrase *ut pictura poesis*, these studies investigate visual description in literature, 'iconic imagery' (it has been called). My own goal is at once less ambitious and more mundane. I try to show how pictures paint poetry, how those for *The Rape of the Lock* explain and expand the text they illustrate.

Besides engraved plates published with the poem in book form, I have treated every kind of illustration based on the poem: engravings designed to be sold separately, as well as drawings,

5. T. S. R. Boase, 'Illustrations of Shakespeare's Plays in the Seventeenth and Eighteenth Centuries', *JWCI*, X (1947), 83–108; W. Moelwyn Merchant, *Shakespeare and the Artist* (1959); Marcia R. Pointon, *Milton and English Art* (Toronto, 1970); Roland M. Frye, *Milton's Visual Imagery and the Traditions of Art* (Princeton, 1978); Ralph Cohen, *The Art of Discrimination Thomson's 'The Seasons' and the Language of Criticism* (Berkeley, 1964), especially Chapt. V.

6. Kester Svendsen, 'John Martin and the Expulsion Scene of *Paradise Lost*', *Studies in English Literature*, I (1961), 64.

watercolours, and oil paintings that were not engraved. Whenever possible I have reproduced the artist's original design along with its engraved version. A fair number of these original illustrations are reproduced for the first time.

My treatment of the illustrations and their artists varies in length. If any illustrations seem to me to lack interest or depth—as many nineteenth-century ones do—I pass over them quickly. (I have ignored anonymous illustrations, all of them hackwork.) If, on the other hand, they merit more extended analysis—those of Fuseli, C. R. Leslie, and Beardsley in particular—then I discuss them more fully. My treatment of these artists is also fuller because I was able to discover more information about their connection with *The Rape of the Lock*. These artists are the ones who, after all, give us what the best illustrators bring to their task: 'a heightened sense of the interrelation of the arts'; they succeed in giving us illustration as illumination.[7]

For a book as short as this, I seem to have imposed on an inordinately large number of scholars, most of them personal friends, and I make grateful acknowledgments to all of them: Miriam J. Benkovitz, G. E. Bentley Jr., Maurice Bloch, Max Browne, the late James Clifford, Judy Egerton, Bernhard Fabian, Ian Fletcher, David Foxon, John B. Friedman, Kenneth Garlick, Robin Hamlyn, Jocelyn Harris, Francis Haskell, Simon Houfe, Keith Maslen, Ronald Paulson, David Piper, Melvin E. Rath, Gert Schiff, John Simmons, Jack Stillinger, Adeline Tintner, Robert Wark, D. H. Weinglass, Stanley Weintraub, and Jerold Ziff. Two generous institutions also deserve my thanks: the Henry E. Huntington Library and Art Gallery for a fellowship and the University of Illinois at Urbana-Champaign (through its Research Board) for various privileges, including the assistance of three excellent graduate students: Susan K. Ahern, Bill J. Rothwell, and Gary Richardson. Finally, I am very grateful to all the owners, whether institutional or private, of the unpublished illustrations and manuscripts who have generously allowed me to print them.

Urbana, Illinois
October 1978

7. Ibid., p. 73

CONTENTS

LIST OF ILLUSTRATIONS

COLOUR PLATES

MONOCHROME PLATES

I
Publication of *The Rape of the Lock* (1714)

The Rape of the Lock. An Heroi-Comical Poem. In Five Canto's. Written by Mr. Pope was published by Bernard Lintot on 4 March 1714. Its title-page, elegantly printed in red and black ink, proclaimed it to be no run-of-the-mill pamphlet; and its author needed no first name or initial to identify him. [*Plate 1*] His 'Pastorals' in 1709 and his *Essay on Criticism* in 1711 had established his reputation; and only three weeks after his new poem appeared he signed an agreement with Lintot for the publication of his *Iliad* translation in a lavish subscription edition. He himself had taken the initiative for this by beginning to enroll subscribers in October of the previous year. It was a nice coincidence that he should simultaneously be translating Homer's grand epic while composing—actually expanding—his own miniature mock-epic.

What he expanded was the earlier version of his poem, entitled 'The Rape of the Locke', published in 1712 by Lintot in the collection *Miscellaneous Poems and Translations*. For this, Lintot had paid him £7.[1] Its fable tells how Belinda, a young woman of fashion, is pursued by the Baron, a London beau, who aspires to possess a lock of her hair; of how during a card party at Hampton Court he cuts off her favourite lock, and despite her anger and a fierce battle refuses to return it to her; and of how, meanwhile, the lock of hair miraculously ascends to the heavens to become a bright comet. Although Joseph Addison, then on friendly terms with Pope, advised him that the 'delicious little thing' was so perfect it should not be altered,[2] Pope disregarded that advice, and determined to expand the poem. While retaining its basic action and witty tone he added the supernatural 'machinery' of Rosicrucian mythology as well as several scenes, including the mordant Cave of Spleen. His

1. John Nichols, *Literary Anecdotes of the Eighteenth Century* (1812–16), VIII, 299.
2. Peter Smithers, *Life of Joseph Addison*, 2nd ed. (Oxford, 1968), p. 242. In Pope's opinion, Addison 'would never alter anything after a poem was once printed. . . .' (Joseph Spence, *Anecdotes*, ed. J. M. Osborn, Oxford, 1966, I, 75).

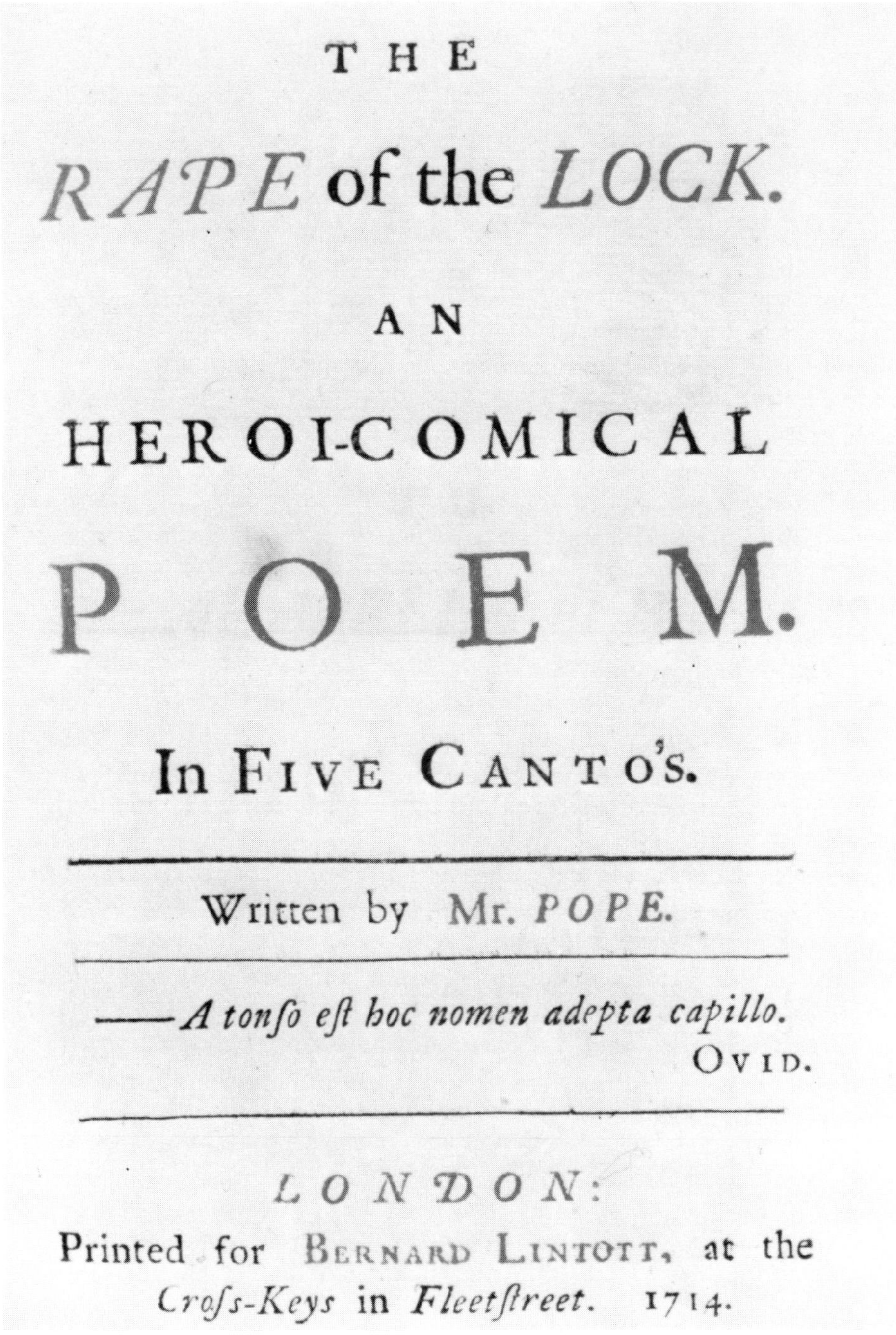

THE

RAPE of the *LOCK*.

AN

HEROI-COMICAL

POEM.

In FIVE CANTO'S.

Written by Mr. *POPE*.

——*A tonſo eſt hoc nomen adepta capillo.*
OVID.

LONDON:
Printed for BERNARD LINTOTT, at the *Croſs-Keys* in *Fleetſtreet*. 1714.

1. Title-page of *The Rape of the Lock* (1714)

additions swelled the poem from two to five cantos, from 334 to 766 lines. In December 1713 he was able to inform Jonathan Swift, 'I have finished the Rape of the Lock'.[3]

By the end of January 1714 Lintot advertised in the *Post Boy* that the poem would be published in a few days, 'complete in 5 Cantos; with 6 Copper Plates'.[4] And on the day of publication, by which time he had paid Pope £15 for the 'Addition',[5] he again advertised the poem in the *Post Boy*: 'with three New Cantos, adorn'd with 6 Copper Plates'.[6] Pope's own phrase for the enlarged and illustrated version, when he told a friend about it, was 'this more solemn edition'.[7]

Its sale was phenomenal—three thousand copies in the first four days, Pope boasted.[8] It remained so popular that in the same year a second and a third edition were printed, in 1715 a fourth edition, in 1718 a fifth, and in 1723 the sixth and last (of the separately printed poem).[9] Although Pope, an inveterate reviser, continued to revise his text in varying degrees, the same copper plates were used in each edition. They became so worn, in fact, that for the fifth edition they had to be retouched.[10]

Although Lintot regarded the copper-plate engravings important enough to stress in his advertisements, Pope—in his surviving correspondence—says nothing of the frontispiece and five other plates or of the artist responsible for them. That artist was Louis Du Guernier, born in Paris in 1687 (hence one year older than Pope). He had emigrated to England about 1708, joining other French artists and engravers who found employment there. As both a draughtsman and an engraver he sometimes performed both functions in the same illustrations, but for *The Rape of the Lock* he only 'invented' the designs, and his countryman Claude Du Bosc, more recently arrived in England, engraved them. In appearance Du Guernier is described by Vertue as being 'of stature rather low than middle size'—which would make him a comfortable companion for Pope—and in temperament 'very obliging, good temper[ed], gentleman-like &

3. *Correspondence*, ed. G. Sherburn (Oxford, 1956), I, 201.
4. Quoted in Twickenham ed., p. 103.
5. Nichols, VIII, 300.
6. Quoted in Twickenham ed., p. 103, n. 2.
7. *Correspondence*, I, 210.
8. *Correspondence*, I, 214.
9. From the ledgers of William Bowyer, the printer employed by Lintot, we know that 1,000 copies were printed for each of the 3rd, 4th, and 6th editions (information from Keith Maslen).
10. R. H. Griffith, *Alexander Pope: A Bibliography* (1922–27; rpt. 1962), I, i, 78.

well beloved by all his acquaintance':[11] obviously a pliable fellow, very willing, apparently, to accept suggestions from the poet and from the publisher for whom he might be working.

We may wonder why Lintot saw fit to embellish the poem with illustrative plates. That he regarded such illustrations as an important part of a book is clear from his own statement in his 1718 edition of Joseph Gardiner's translation of Rapin's poem on gardens. There the Bookseller tells the Reader that although prevented from praising the translation, 'yet I have shewn my own Esteem of it, by the Care I have taken in the printing of this Edition, and the Expence I have been at in adorning it; which was the highest Expression of Gratitude that would be accepted by Mr. Gardiner, from his and the Reader's Humble Servant, Bernard Lintot'.[12] We may assume that as a mark of his high opinion of Pope's poem he was willing to lay out the extra expense of providing illustrations. He also distinguished the 1714 pamphlet in another way: it was evidently the first octavo book of English verse to be decorated with engraved headpieces, a tailpiece, and an ornamental initial letter, decorations that had previously been reserved mainly for stately folios.[13]

How frequently were single poems illustrated? Thanks to David Foxon's monumental catalogue, we can with some certainty calculate that in the decade leading up to the publication of *The Rape of the Lock*—from 1704 to 1713—among more than eleven hundred separately printed poems, only six were illustrated with plates. Six out of eleven hundred!—clearly, an illustrated poem at this time was uncommon. This paucity is also apparent when we calculate that in the decade following 1714—that is, from 1715 to 1724—of a total of almost seven hundred and fifty separately printed poems, only six were illustrated.[14] In summary, from 1704 to 1724 (omitting the year 1714) out of almost two thousand separately published poems only twelve contained illustrative plates.

The single year 1714 was extraordinary for illustrated poems, and especially for Du Guernier's work: Samuel Garth's enormously popular *Dispensary*, which in its earlier editions (published by John

11. George Vertue, *Notebooks* (Oxford, 1930–42), II, 38. The fullest treatment of Du Guernier is in Hanns Hammelmann, *Book Illustrators in Eighteenth-Century England*, ed. T. S. R. Boase (New Haven and London, 1975), pp. 47–8.

12. *Rapin Of Gardens A Latin Poem*. . . . English'd by Mr. [Joseph] Gardiner, 2nd ed.

13. D. F. Foxon, *Pope and the Early Eighteenth-Century Book-Trade* (typescript, 1976), f. 51.

14. D. F. Foxon, *English Verse 1701–1750: A Catalogue of Separately Printed Poems* (Cambridge, 1975), II. From 1704 to 1724 about fifty-five poems were published with only frontispieces, many of them illustrative, but I am not counting them in my statistics.

Nutt) contained only an unsigned architectural frontispiece, was issued in June by Tonson in its seventh edition with a frontispiece and six plates, all by Du Guernier; John Gay's *Shepherd's Week*, nominally published by Burleigh but actually by Tonson, had a frontispiece and six plates also by Du Guernier;[15] and Pope's poem had a frontispiece and five plates. Also published that spring, by Edmund Curll, was a religious work by Edward Young illustrated by a different artist.

The poems by Garth, Gay, and Pope, all illustrated by Du Guernier, are also linked through their common literary genre: two of them mock-heroical and one mock-pastoral. What might have suggested to these rival publishers that they should illustrate the poems was the publication two years previously of an illustrated edition of the mock-heroic poem *par excellence*, the one that was most prominently the model of the genre. Boileau's *Le Lutrin*,[16] first translated by John Ozell in 1708 and issued then with only a frontispiece, contained five additional full plates in its second edition (1712), when published in the first volume of Boileau's works. But the third edition, issued separately with these plates, was published in 1714. It seems to have been the *annus mirabilis* for illustrated poems: five in one year compared with six in the previous decade and six in the following one.

Is there any reason, one wonders, to think that Pope had any control over the illustrations that accompanied his poem? If he did, then the plates can be assumed to embody his ideas. There is some evidence, though not directly concerned with *The Rape of the Lock*, that supports this assumption. Three weeks after the poem was published his agreement for publication of his *Iliad* translation provided that he was to have a special quarto edition for subscribers, with headpieces, tailpieces, and initial letters engraved on copper 'in such manner and by such Graver as the said Alexander Pope shall direct and appoint'.[17] Since Lintot conceded this privilege for Pope's Homer, it is likely that he had made a similar concession for *The Rape*

15. Gay's biographer writes that *The Shepherd's Week* was 'pleasantly equipped with engravings' (William Henry Irving, *John Gay Favorite of the Wits*, Durham, N.C., 1940, p. 85); and his most recent editor does not even mention the illustrations, though he provides thirty pages of commentary on the poem (*Poetry and Prose*, ed. Vinton A. Dearing, Oxford, 1974).

16. In the 1675 edition of Boileau's *Œuvres diverses*, *Le Lutrin* had only a single illustrative plate (Paris: Denys Thierry, facing p. 195), and in the 1683 edition an illustrative frontispiece (Paris: Claude Barbin).

17. Dated 23 March 1713 [O.S.]; in Pope, *Poems*, Twickenham Edition, X (1967), p. 606.

of the Lock, and that Pope, who from his earliest dealings with the printing trade took an 'active and innovatory' interest in the physical appearance of his printed works,[18] would have claimed the privilege of choosing the illustrator and 'editing' the illustrations. Furthermore, in 1713—the year in which the illustrations for his poem were being invented and engraved—Pope himself was engaged in an intensive study and practice of painting under the tutelage of Charles Jervas; 'so entirely immersed in the designing art. . .', he tells John Caryll, 'that I have not heard a rhyme of my own gingle this long time'.[19]

It is hard to imagine that he would take no interest in the designs then being made to accompany the rhyme of his own gingle. He once told Spence that he could not say which gave most pleasure, poetry or painting, because both were so extremely pleasing to him.[20] In his epistle to Jervas, written about this time, he celebrates their 'love of Sister-arts'; and one couplet can literally be applied to the illustrative plates that faced his poetry:

> How oft' our slowly-growing works impart,
> While images reflect from art to art?[21]

Could he have resisted taking an interest in the engraved images (by Du Guernier) that reflected his own verse?

In allowing Pope to choose the illustrator and illustrations for the *Iliad* translation (and probably for *The Rape of the Lock* as well) Lintot was conforming to publishers' practice in his day. Other publishers, we know, relied on their authors to choose both illustrators and illustrations.

John Gay was evidently accustomed to dealing with his illustrators. 'I am about to publish a collection of Fables entirely of my own invention. . .' he wrote to a friend (in 1726); 'they consist of fifty, and I am oblig'd to Mr Kent & Wootton for the Designs of the Plates. The Work is begun to be printed, and is delay'd only upon account of the Gravers, who are neither very good or expeditious.'[22] He not only chose the artists but followed the slow progress of the engravers.

18. Foxon typescript, f. 13.
19. *Correspondence*, I, 189. Evidence of Pope's practice as a draughtsman is the collection of seventeen of his drawings in the inventory of his estate (*Notes and Queries*, 6th series, V, 1882, p. 363). For Pope's interest in painting and his practice of it, see Norman Ault, *New Light on Pope* (1949), Chapt. V.
20. Spence, *Anecdotes*, I. 46.
21. Pope, *Poems*, Twickenham Edition, VI (1964), p. 156.
22. Gay, *Letters*, ed. C. F. Burgess (Oxford, 1966), pp. 62–3.

Swift was also involved with his illustrators. The fifth edition of *A Tale of a Tub* in 1710 was embellished with eight engraved plates illustrating the text; and although the evidence is uncertain Swift 'probably discussed their subjects, and perhaps some of their details' with his friend Sir Andrew Fontaine, to whom the drawings were sent for alteration.[23] More definite evidence of Swift's interest in illustrations appears many years later, in December 1727, when his publisher Benjamin Motte wished to bring out an illustrated edition of *Gulliver's Travels* a year after its initial publication. He asked Swift (who was in Dublin) to suggest which episodes in the book should be illustrated. In his long, detailed reply, Swift listed eight scenes from Book I: 'Some of these,' he continued, 'seem to be the fittest to be represented, and perhaps two adventures may be sometimes put in one Print.' He then listed seven scenes from Book II, only one from Book III (the floating island), and six from Book IV. 'I can think of no more,' he concluded; 'But Mr Gay will advise you and carry you to Mr Wotton, and some other skilfull people.'[24] Not only did Swift suggest the scenes to be illustrated but he recommended John Wootton, the well known animal painter. (Wootton could obviously be trusted to portray the Houyhnhnms in Book IV.)

Motte's request had not been an idle one. In the next edition (the third) of *Gulliver's Travels*, dated 1727 but published early in 1728, four illustrative plates were inserted. This was the first duodecimo edition of the book. (Publishers frequently added illustrations to stimulate sales of a new edition.) The four plates contain ten scenes from the book, and of these, five conform to Swift's suggestions, including the floating island, the only scene to occupy an entire plate. (Each of the other plates has three scenes.) Neither the artist's nor the engraver's name (perhaps they were the same man) appears on the plates. These illustrations were repeated when the duodecimo edition was reissued.[25] The episode is instructive in showing how the author, even at a distance from his publisher, was consulted, and his advice followed, when his work was to be illustrated.

When Voltaire published his epic *La Henriade* in London in 1728

23. *A Tale of a Tub*, ed. A. C. Guthkelch and D. N. Smith, 2nd ed. (Oxford, 1958), pp. xxv–xxviii.

24. Swift, *Correspondence*, ed. H. Williams (Oxford, 1963–65), III, 257–8; also 263.

25. Herman Teerink and Arthur H. Scouten, *A Bibliography of the Writings of Jonathan Swift*, 2nd ed. (Philadelphia, 1963), Nos. 294, 295, 304, 305. The illustrations are listed (but not discussed) in David S. Lenfest, 'A Checklist of Illustrated Editions of "Gulliver's Travels", 1727–1914', *Papers of the Bibliographical Society of America*, 62 (1968), 87.

he personally and precisely supervised its elaborate illustrations—ten large plates and ten vignettes.[26] He himself had chosen the illustrators and scenes to be drawn,[27] and he had had the plates engraved in Paris previously so that they were ready even before he had completed his new version of the poem.[28] The most explicit evidence of an author's (or editor's) control of his illustrations can be seen in the contract drawn by Sir Thomas Hanmer, who (in 1740) hired Francis Hayman to illustrate his edition of Shakespeare. The subject of each drawing was to be chosen by Hanmer, and each drawing was subject to his approval.[29]

In regard to *The Rape of the Lock* and its illustrations in 1714, several remarkable coincidences connect Pope and Du Guernier with Gay, who was the poet's intimate and devoted friend. In October 1713 Pope writes to him, 'I am deeply ingaged in Poetry, the particulars whereof shall be deferr'd till we meet.'[30] At the same time Gay was completing his own poem *The Shepherd's Week*, which, Samuel Johnson later wrote, Pope was supposed to have 'incited' him to write.[31] Both poets, then, were completing mock-poems; both poems were published only one month apart the following spring (by different publishers); and, most strikingly, both poems were illustrated by Du Guernier. It is difficult to avoid the assumption that Du Guernier, working on Pope's and Gay's poems at the same time, conferred with them about the illustrations, and that his illustrations to *The Rape of the Lock* were suggested or accepted by Pope.

26. Juliette Rigal, 'L'Iconographie de la *Henriade* au XVIII siècle', *Studies on Voltaire and the Eighteenth Century*, ed. Th. Besterman, XXXII (Geneva, 1965), p. 25.
27. *Correspondence*, ed. Th. Besterman (Geneva, 1953), I, 164–7.
28. Gay, *Letters*, p. 63.
29. Manuscript in copy of Shakespeare's *Works* (1744), ed. Thomas Hanmer, in the Folger Shakespeare Library. The agreement is printed in facsimile in *Shakespeare Quarterly*, IV (1953), facing p. 286 and transcribed in ibid., IX (1958), 141. See also Marcia Allentuck, 'Sir Thomas Hanmer Instructs Francis Hayman: An Editor's Notes to his Illustrator (1744)', ibid., XXVII (1976), 288–315.
30. *Correspondence*, I, 195.
31. *Lives of the Poets*, ed. G. B. Hill (Oxford, 1905), II, 269.

II
The Illustrations of 1714

Contemporary readers of the poem, in all its editions from 1714 to 1723, were confronted by the copper-plate engraving that faced the title page and by the engravings that faced the opening lines of the five cantos. These illustrations, which readers unavoidably had to scan, thus contributed to what can be called their reading experience.

As graphic evidence the plates illustrate explicit characters, settings, and objects, delineating in a literal sense much of what the text sets forth. But they also point up the most important scene in each canto, for the poem as a whole contains no fewer than fifteen 'tableaux', and from these, only five were chosen to be illustrated. This choice was almost surely made by Lintot, who would have consulted Pope about it. And more important, the plates add to the poem information that is *not* in the text. If we concede the probability that Pope controlled the illustrations then we may regard this information as his addition to his own text. The plates supplement the text not by contradicting or nullifying but rather by supplementing it; they tell us more about the settings, characterizations, and action (or plot).

The illustration for Canto I [*Plate 2*], the opening scene of the poem, displays the heroine Belinda asleep, while Ariel, her guardian sylph, hovers above and warns her in a dream that she must beware of an approaching 'dread Event'. Judged by this design, Du Guernier would seem to be an inferior artist. Good drawings, however, can be spoiled by a poor engraver, and the illustrations that Du Guernier both drew and engraved in other books are not so clumsy as this. His frontispiece for Gay's *The Shepherd's Week*, for example, drawn at about the same time as his illustrations for Pope's poem, is graceful and animated; and following a venerable tradition, he drew a self portrait (in the lower right), a kind of personal signature. [*Plate 3*]

As Belinda completes her morning dream her lapdog Shock

2. Louis Du Guernier and Claude Du Bosc, Canto I of *The Rape of the Lock* (1714), engraving

3. Louis Du Guernier, frontispiece to *The Shepherd's Week* (1714) by John Gay, engraving

'Leapt up, and wak'd his Mistress with his Tongue'.[1] Her room is furnished with objects mentioned in the poem. The 'white Curtains' that in Pope's verse surround her bed are transformed here to what looks like a military tent—which may be a satiric comment on the heroic trivialities that follow. (It may also be a residue from the designs that Du Guernier was then drawing for the battles of the Duke of Marlborough.) On Belinda's dressing-table we see the mirror and a few of the objects catalogued at the end of the canto when she performs 'the sacred Rites of Pride'. The small square object may be a Bible (in the much discussed line: 'Puffs, Powders, Patches, Bibles, Billet-doux').[2] Visible beneath the skirt of the dressing-table is an ambiguous form: either a leg, ending in a claw-and-ball foot, of the stool on which Belinda will sit while worshipping 'the *Cosmetic* Pow'rs'—or the leg and cloven hoof of a satyr. If the latter, then the illustration has introduced a pagan symbol unmentioned in the text, a hint of the poem's erotic underpinning.

The plate for Canto II shows us Belinda 'Lanch'd on the Bosom of the Silver *Thames*' in the 'painted Vessel'. [*Plate 4*] Pope's text also tells us that the vessel has sails. Although the customary passenger boat on the Thames was a barge—a common word not used by Pope—furnished with a rowing crew but without sails, boats like the one depicted by Du Guernier did occasionally ply the waters.[3] Pope's decision to introduce this type of boat could not be disregarded by his illustrator, for the sylphs (in the verse) cluster around the 'sails' and 'gilded Mast' of the vessel. Mediating between the poetic image and the Thames reality Du Guernier designed a passenger boat with both rowers and a sail. He also distorted the shape of the little boat—which should be much longer—so that it resembles the Ship of Fools often illustrated in Renaissance books. Ariel instructing his 'lucid

1. For the erotic connotations of lapdogs see Donald Posner, *Watteau: A Lady at her Toilet* (1973), Chapt. 6. Lapdogs as erotic partners for women occur in at least two contemporary English satires: [Robert Gould], *Love Given O're: or, a Satyr against the Pride, Lust, and Inconstancy &c. of Woman* (1682), p. 5; [Richard Ames], *The Folly of Love; or, an Essay upon Satyr against Woman* (1691), p. 8. At the same time Pope did not make explicit use of his Rosicrucian source in its notion that sylphs sometimes took on the shapes of dogs 'and so address themselves to the wanton frailty of *women*; who are affrighted at a Lovely *Sylphide*, but not at a *Dog*, or *Munkey*. I could tell you many Tales of your little *Bolognian Dogs*, and certain pretty *Ladyes* in the World. . . .' (Abbé de Montfaucon de Villars, *The Count of Gabalis*, transl., 1680, pp. 156 7).
2. Appendix F in Twickenham Edition of *The Rape of the Lock*.
3. Maynard Mack, *The Garden and the City* (Toronto, 1969), plate 5. It was more common for freight-carrying boats to use sails.

4. Louis Du Guernier and Claude Du Bosc, Canto II of *The Rape of the Lock* (1714), engraving

5. Louis Du Guernier and Claude Du Bosc, Canto III of *The Rape of the Lock* (1714), engraving

Squadrons' how to protect Belinda: what a challenge to any illustrator! At least Du Guernier has shrunk the gross and lumpish Ariel of the first plate to a delicate creature closer to Pope's description. In the distance, past the bend in the river, looms the East Front of the Christopher Wren palace at Hampton Court, where (the poem tells us) 'great *Anna*' sometimes counsel took and sometimes tea. Queen Anne actually preferred to live at her other palaces, and came to Hampton Court only for brief visits or to attend meetings of her Privy Council.[4]

In Canto III of the poem the rape of Belinda's lock takes place in a drawing-room of the palace, after the card game, while she bends her head over the 'fragrant Steams' of coffee and tea; but Du Guernier's illustration sets the scene outside. [*Plate 5*] He adapts that part of the palace called the Ionic Colonnade to create a setting more appropriate to the mock-heroic epic: the massive pillars (their pedestals enormously enlarged), the palace façade, the huge billowing drapery, and the distant trees. The characters are caught in the exaggerated rhetorical postures of stage-actors. In fact, Du Guernier designed frontispieces for stage drama, particularly Shakespeare's in the Tonson 1714 reprint of Nicholas Rowe's 1709 edition.

Most prominent in this theatrical tableau, the Baron triumphantly holds aloft his 'glorious Prize' and, in his other hand, the 'fatal Engine' that had severed it. Belinda, visibly suffering from 'Rage, Resentment and Despair', is being consoled by a man and a woman. The man's cane identifies him as Sir Plume, and the woman must be the 'fierce *Thalestris*', whose beau he is. Yet these two characters do not appear in the poem until the next canto, where Sir Plume is Belinda's champion and Thalestris her confidante. Their presence here amplifies their dramatic function and motivates their later actions. It is doubtful that the illustrator would have taken this liberty without the poet's suggestion or approval. As to the woman sitting at the small, triangular card-table: who is she? In the poem Clarissa—as 'Ladies in Romance assist their Knight'—simply hands the Baron the scissors, the 'two-edg'd Weapon from her shining Case'. (The scissors-case may be on the table, whose surface is somewhat indistinct.) She is not mentioned again until Canto V. In this illustration she displays a further dramatic action by disdainfully

4. Daniel Lysons, *Historical Account of those Parishes . . . not described in the Environs of London* (1800), p. 66; Ernest Law, *History of Hampton Court Palace* (1891), III, 170, 171.

looking away from the mock-tragedy she has helped to bring about. Her dramatic role is thus expanded and sharpened.[5] The dog Shock barks angrily (in normal canine fashion) at the villain who has disturbed his mistress. After his prominent role in Canto I, where he awakens Belinda, he no longer plays a part in the poem; the illustration restores him to the cast of active characters in Canto III.

In this plate, then, the illustrator—probably at the poet's direction—has expanded the scenario of the poem's verbal text. By placing Sir Plume and Thalestris at the scene of the lock's rape he strengthens their subsequent actions; and by depicting Clarissa after the poet has abandoned her (until later), he has made the plot of the poem more coherent and forceful. Three years later, in the 1717 version of the poem, Pope himself emphasized the importance of the 'grave *Clarissa*' by adding a long speech for her in Canto V.

This is the only one of the six plates that does not depict any of the sylphs or gnomes, and that omission is subtly justified by the text. For just as the Baron prepares to sever Belinda's lock, Ariel views an 'Earthly Lover lurking at her Heart':

Amaz'd, confus'd, he found his Pow'r expir'd,
Resign'd to Fate, and with a Sigh retir'd.—

presumably followed by his legions of sylphs, all of them powerless to avert the tragedy.

In Canto IV the gnome Umbriel descends to 'the gloomy Cave of *Spleen*'. [*Plate 6*] Described in the poem as a 'dismal Dome', the 'Grotto' where Spleen resides, it is drawn by Du Guernier to suggest a spacious cave extending into the distance, with a darkly ominous overhang in the foreground. This plate is particularly valuable for solving an ambiguity in the text. The wayward Queen of this domain has two 'Handmaids', Ill-nature and Affectation; and as she sighs on her pensive bed she also has '*Pain* at her side, and *Languor* at her Head'. In the fourth edition of the poem (1715) Pope substituted Megrim for Languor, but the uncertainty still remains: are Pain and Languor (or Megrim) simple afflictions or are they allegorical personifications, like the two named Handmaids?[6] Du Guernier's illustration firmly solves the crux by depicting only the two Handmaids, thus interpreting Pain and Languor as simple ills suffered by the Queen. In doing this he was presumably carrying out

5. See also John Trimble, 'Clarissa's Role in *The Rape of the Lock*', *Texas Studies in Literature and Language*, 15 (1974), 673–91.
6. Tillotson calls them 'allegorical figures' (Twickenham Edition, p. 185, n. 24).

6. Louis Du Guernier and Claude Du Bosc, Canto IV of *The Rape of the Lock* (1714), engraving

7. Louis Du Guernier and Claude Du Bosc, Canto V of *The Rape of the Lock* (1714), engraving

Pope's intention. Later illustrators, as we shall see, departed from this interpretation.

For Ill-nature, an '*ancient Maid*', Du Guernier invented a creature even more witch-like than the witches he designed for Tonson's 1714 edition of *Macbeth*—with sharp, gnarled features, a long nose that threatens to meet her jutting chin, and skinny, pendulous breasts. But he must have been challenged by the couplet:

With store of Pray'rs, for Mornings, Nights, and Noons,
Her Hand is fill'd; her Bosom with Lampoons.

Are the store of prayers and the lampoons of Ill-nature to be taken literally or figuratively? In Du Guernier's illustration, sheets of paper are dimly visible in Ill-nature's left hand. Her store of prayers visible, her rancorous lampoons hidden in her breast, she is a consummate hypocrite.

In the 'constant *Vapour*. . . Strange Phantoms' rise, the poem states. The plate shows a few 'glaring Fiends, and Snakes on rolling Spires', and on the ground two crustaceans, perhaps from the 'Unnumber'd Throngs . . . Of Bodies chang'd to various Forms by *Spleen*'. More specifically, Du Guernier depicts (in the lower left) a group of five objects described in the passage: 'living *Teapots* stand, one Arm held out,/One bent; the Handle this, and that the Spout'. Although he omits walking pipkin, sighing jar, and pregnant men—'Men prove with Child, as pow'rful Fancy works'—[7] he does illustrate the suggestive line: 'Maids turn'd Bottels, call aloud for Corks', for we see two bottles already corked—with men's heads! What in the text can be taken as pure fancy or fantasy is in the illustration rendered unmistakably concrete and earthly. In that group, the flat, box-like object, showing the faint outlines of a figure, is puzzling: it may be one of the 'Visions of expiring Maids' or 'Angels in Machines' or the talking 'Goose[berry]pye'.

In the left foreground the gnome Umbriel is departing from the Cave. In characterizing this creature as roaming the earth in search of mischief Pope differs from his source. To the Rosicrucians the gnomes, a people of small stature, are guardians of treasures; they are 'Ingenious, Friends of Men, and easie to be commanded'.[8] Du

7. The annotation to this line in the Twickenham Edition cites a contemporary clergyman and Dryden's *Wild Gallant*. More relevant, perhaps, is that in the Renaissance this notion was classified as a type of madness (Hyacinthe Brabaut, 'Les traitements burlesques de la folie aux xvie et xviie siècles', *Folie et déraison à la Renaissance*, Bruxelles, 1976, p. 83 and Fig. 13).

8. Villars, p. 29.

Guernier obviously followed Pope's text in drawing the sooty gnome as he departs for the world above to do his mischief. In his right hand he carries the vial filled with fears, sorrows, griefs, and tears; and in his other hand, not clearly shown, he carries the bag of sighs, sobs, passions, and war of tongues. When released, the contents of the vial and bag will provide Belinda with ammunition for her emotional outburst.

In Canto V Belinda and her cohorts battle against the Baron and his in an attempt to regain her ravished lock of hair. [*Plate 7*] In the foreground the chair thrown over by the violence of the struggle is a motif copied from the *Hamlet* illustration by François Boitard in Rowe's 1709 edition of Shakespeare, though when Du Guernier revised that plate for the 1714 reprint he eliminated the chair. (One might say that he carried the chair from *Hamlet* to *The Rape of the Lock*.) The illustration also serves to reinforce the metaphors in the text. For when Belinda subdues the Baron with a 'Charge of *Snuff*', the violence of his sneeze has actually hurled him to the ground—a literal interpretation, thus, of his defiant 'Boast not my Fall (he cry'd) insulting Foe'. He lies half recumbent on the floor, his hat beside him while she points her 'deadly *Bodkin*' toward his heart. The woman at our right, who is about to strike the man with her fist, must be Belinda's friend, the fierce Thalestris; and on the far left the woman thrown to the floor but managing to hold on to her fan must be Clarissa, brought down by Sir Plume, while in the background a woman (Chloe) looks in his direction. Their actions illustrate the lines:

> As bold Sir *Plume* has drawn *Clarissa* down,
> *Chloe* stept in, and kill'd him with a Frown;
> She smil'd to see the doughty Hero slain,
> But at her Smile, the Beau reviv'd again.

The revived Sir Plume is actually helping Clarissa to her feet, a gallant action not mentioned in the text. It seems unlikely that Du Guernier would add this without Pope's suggestion or permission.

Above the embattled humans 'Triumphant *Umbriel* on a Sconce's Height' looks on, accompanied by two gnomes. Although the text sets the battle scene under 'vaulted Roofs' and a 'high Dome', no room in the Wren palace at Hampton Court fits this description, nor does the illustration. Instead, Du Guernier designed an atrium-like room with a balustraded aperture to reveal the comet in the sky. The witty irony of the poem's conclusion is reinforced, for even as the battle rages the ravished lock has been transformed to add 'new

Glory to the shining Sphere': the heroic contest is thus a futile exercise.

The frontispiece is the most provocative of all the plates. [*Plate 8*] Although the first to meet the reader's eye it does not yield up its meaning until after he has read the entire poem. While each canto has a simple illustrative plate, the frontispiece is synoptic, an overture announcing the themes to be developed. In effect, its mixture of symbols and allusions piques the reader's curiosity.

Immediately recognizable, however, is the architectural backdrop, the East Front of the Wren palace, with (on the left) one of the two ornamental, decorated urns that stood in front of the façade. A group of putti, meant to suggest sylphs, float in air; one points up to the comet in the sky, directing the viewer's eye to the apotheosis of both the raped lock and the poet's fame. Another putto drops, by accident or design, a cascade of playing cards. Centered in the main cluster of figures, the seated female looking into the mirror suggests a Venus-like Belinda adoring the cosmetic powers; her mirror is held before her by a putto, a common motif exemplified in Titian's and Velasquez's famous paintings of Venus adoring her reflection in the mirror held by Cupid.[9] Her exposed leg is a highly erotic gesture. Biblical and later moralists link the mirror and exposed leg as marks of a lustful woman;[10] and since the figure is obviously intended to represent Belinda, the illustration emphasizes with a more serious tone the coquetry so elegantly mocked by Pope's couplets.

On the far right stands a large putto, the only one of the terrestrial group clearly equipped with wings. He clutches a vial—possibly an allusion to Umbriel's from the Cave of Spleen—while he puts his other hand to his mouth in a gesture possibly denoting silence. Between him and the mirror, hardly visible in the shadow, lurks a mysterious, demonic figure. If the mirror symbolizes vanity or pride, as it usually does, then this figure may embody the warning (in Canto I) that nymphs 'too conscious of their Face' are predestined 'to the *Gnomes* Embrace'. The theme of female vanity is further emphasized in the oversized putto (in the left foreground) wearing a pair of high-heeled shoes similar to Belinda's in Plate I. Lying on the ground before him are several objects of female vanity, including a large jewel casket; all these illustrate what is clearly mentioned in the poem. But how are we to interpret the bestial figure seated in the

9. See also Heinrich Schwarz, 'The Mirror in Art', *Art Quarterly*, 15 (1952), 97–118.

10. John B. Friedman, 'L'Iconographie de Vénus et de son miroir à la fin du moyen âge', *L'Érotisme au moyen âge*, ed. B. Roy (Montreal, 1977), pp. 66–9.

8. Louis Du Guernier and Claude Du Bosc, frontispiece to *The Rape of the Lock* (1714), engraving

right foreground: a satyr with cloven hooves, goatlike hairy haunches, and pointed ears, and holding an animal mask in front of his face?

The satyr with mask identifies the poem's satiric purpose, of course,[11] but more graphically it emphasizes the poem's erotic content. Noted by commentators soon after its publication and very frequently in our post-Freudian era, the erotic element is only indirectly and obliquely hinted in the text. As Geoffrey Tillotson has written: 'because of the very perfection of the surface of the poem, it is possible to read the *Rape of the Lock* . . . without realizing that below the exquisite scintillation an ingenious obscenity is sometimes curling and uncurling itself.'[12] Nowhere in the poem is there any mention of a satyr, a pagan figure (one of Pope's contemporaries wrote) used by the ancients for 'a passion too gross to be named', representing 'the vice and brutality of the sensual appetite'.[13] We can recall the leg, possibly a satyr's, visible under Belinda's dressing-table. The emphasis on the erotic element in these plates can be matched by Pope's own emphasis when he had enlarged the 1712 version of the poem for this 1714 edition. In the latter edition we are told that since a sylph can protect a maiden only if she is 'fair and chaste', Ariel's power to protect Belinda expires when he views 'in spite of all her Art,/An Earthly Lover lurking at her Heart'. The implication is clear.

The most blatantly sexual couplet Pope had transferred from Thalestris's angry speech in 1712 to Belinda's lament when she addresses the Baron:

> Oh hadst thou, Cruel! been content to seize
> Hairs less in sight, or any Hairs but these!

What else could this mean except pubic hairs, and who else could seize them except an earthly lover? This smutty allusion was recognized at the time not only by an unfriendly reader like Charles

11. Elizabethan satires frequently used satyrs as title-page ornaments. Although the false etymology of 'satire' as derived from *satyrus* (rather than *satura*, a medley) had been exposed in the early seventeenth century, the satyr continued to be used with this symbolic meaning. In 1726 Hogarth put a large satyr in the frontispiece of his set of engravings for Butler's *Hudibras* (1663).

12. *On the Poetry of Pope*, 2nd ed. (Oxford, 1950), p. 157.

13. Lady Mary Wortley Montagu in *Essays and Poems*, ed. R. Halsband and I. Grundy (Oxford, 1977), p. 386. Pope's own use of 'satyr' in this sense is remarkably rare—only twice in all his works: in his 'Pastorals' (1709) and unavoidably in a translation from Ovid (1712) (Twickenham Edition, *Poems*, I, ed. E. Audra and A. Williams, 1961, pp. 75, 378).

Gildon,[14] but by Pope's good friend John Gay, in a letter to their friend Charles Ford about Arabella Fermor, the dedicatee of Pope's poem, only a few months after its publication. 'There are shades at Bingfield, Mrs Fermor is not very distant from thence; make a visit to Pope and Parnelle, and while they are making a Grecian Campaign, do you as Æneas did before you meet your Venus in a Wood, he knew her by her *Locks* and so may you—but as you are a man of Honour & Modesty—think not of Hairs less in sight or any Hairs but these.'[15]

When Pope's friend Thomas Parnell paid tribute (in 1722) to his genius in the verse 'To Mr. Pope' he refers to satyrs, in the phrase 'satyr train'.[16] In 1745, a year after Pope's death, one of his champions more emphatically described the frontispiece as a tribute to his art and imagination. In *The Rape of the Lock*, the anonymous poet writes, Pope created a fairy land filled with sylphs and gnomes; in the sky the comet of Belinda's hair

> Flames o'er the Night. Behind, a Satyr grins,
> And, jocund, holds a Glass, reflecting, fair,
> Hoops, Crosses, Mattadores; Beaux, Shocks, and Belles,
> Promiscuously whimsical and gay.[17]

The grinning satyr emerges not from Pope's lines but from the frontispiece.

All of Du Guernier's illustrations in other books, with very few exceptions, are literal, unimaginative, uncomplicated. (One exception, similar to his frontispiece for *The Rape of the Lock*, is that for the 1714 edition of *The Dispensary*.) Yet Pope himself, when he painted a frontispiece for the *Essay on Man* many years later, designed this kind of thematic composition, an overture of symbolic elements from the poem it introduces.[18]

14. *A New Rehearsal, or Bays the Younger* (1714), pp. 43–4. See also John Oldmixon, *The Catholic Poet* (1716), p. 1; John Dennis, 'Remarks on Mr. Pope's *Rape of the Lock*', [1714] 1728, *Critical Works*, ed. E. N. Hooker (Baltimore, 1939–43), II, 347–8; James Ralph, *Sawney* (1728), pp. 9–10; *The Life of Alexander Pope. . . .* (1744), pp. 16–17; William Ayre, *Memoires of the Life and Writings of Alexander Pope* (1745), I, 41; Thomas Tyers, *An Historical Rhapsody On Mr. Pope* (1782), pp. 37–8.
15. Gay, *Letters*, p. 11.
16. For Parnell's verse, see p. 37 below. In Earl Wasserman's view, 'No doubt Parnell took his clue from the illustration in the 1714 edition of *The Rape of the Lock* which shows cupids attending Belinda at her toilet and satyrs [*sic*] at the edge of the group' ('The Limits of Allusion in *The Rape of the Lock*', *JEGP*, 65, 1966, p. 441, n. 41).
17. *A Plan of Mr. Pope's Garden . . . To which is added a Character of all his Writings* (1745), p. 25.
18. *Essay on Man*, ed. M. Mack, Twickenham Edition, III.i (1950), frontispiece.

Pope's connection with the frontispiece of *The Rape of the Lock* is strengthened when we examine the 1717 version of the poem included in the quarto edition of his *Works*, also published by Lintot. Du Guernier's octavo frontispiece (and plates) could not of course be used, nor could Du Guernier himself, dead since the previous year. Instead, the accomplished and prolific artist-engraver Simon Gribelin, who had supplied two ornamental headpieces for the 1714 edition of the poem, designed and engraved a headpiece for each poem in this volume. For *The Rape of the Lock* he adapted the Du Guernier 1714 frontispiece. [*Plate 9*] We can also be more certain that Pope controlled the choice and use of this illustration, for he instructed Lintot most carefully on how this volume should be designed. 'I desire, for fear of mistakes,' he writes, 'that you will cause the space for the initial letter to the Dedication to The Rape of the Lock to be made of the size of those in Trapp's Prælectiones. Only a small ornament at the top of that leaf, not so large as four lines breadth. The rest as I told you before.'[19] He could hardly be more finicky. 'The rest as I told you before' suggests that he was accustomed to giving instructions to Lintot, and that Lintot was accustomed to following them.

In the illustration itself the center medallion shows, again, the East Front of the Wren palace at Hampton Court, more accurately drawn than in Du Guernier's plates. As in the 1714 frontispiece a sylph directs our attention to the heavenly body—a star this time rather than a comet. Other sylphs are engaged in various activities mentioned in the poem: playing cards (around a square rather than a triangular table), gazing into a mirror (vanity), waving a fan (coquetry), holding up a mask (pretence). And in place of the smallish satyr of the 1714 frontispiece two large satyrs dominate the design, on either side of the medallion, each peering at the scene through one mask and holding another (comedy and tragedy). The helmet and wings at the top of the cartouche symbolize Mercury, the Roman divinity who protects traders and thieves. He is invoked here to protect the ravisher of the lock of hair.

This is not the only poem of Pope's whose illustrations have been scrutinized for their relation to his text. In a recent essay on the *Dunciad* illustrations—of which there are seven headpieces—a scholar comes to the conclusion that because of Pope's great care in the printing of his works 'we must assume that he is responsible for

19. *Correspondence*, I, 394.

9. Simon Gribelin, Pope's *Works* Vol. I (1717), engraving

all the *Dunciad* prints. . . he must have directed the artists or, at the very least, approved the designs submitted to him'.[20] These assumptions hold true for *The Rape of the Lock*.

20. Elias F. Mengel, Jr., 'The *Dunciad* Illustrations', *Eighteenth-Century Studies*, 7 (1973–74), 161. See also Benjamin Boyce, 'Baroque into Satire: Pope's Frontispiece for the *Essay on Man*', *Criticism*, IV (1962), 14–27.

III
Later Eighteenth-Century Illustrations

Later illustrations for *The Rape of the Lock* do not raise the intricate question of whether Pope collaborated with the artists by supervising or approving their designs. Instead we wonder to what extent these illustrations reflect contemporary critical attitudes toward the poem; and, in addition, how the illustrations reflect the artist's own vision as well as the painterly attitudes and techniques of his time. Thus, although the text of the poem remained unchanged, its meaning and significance were continually being altered by the taste and sensibility of the critics who read it and by the artists who translated—or transmuted it—into the iconographic idiom of their time.

As early as 1717 William Hogarth, then beginning his career, may have designed a scene from the poem. [*Plate 10*] As described by Joseph Warton (in 1797) this 'engraving of Sir Plume, with seven other figures, by Hogarth, was executed on the lid of a gold snuff box, and presented to one of the parties concerned. . . .'[1] Hogarth's recent biographer writes, more cautiously, that the engraving is probably, though not certainly, by the artist.[2] The scene is Canto IV of the poem: in the foreground Sir Plume, identified by his cane and snuffbox, is trying to persuade the Baron to surrender the lock of hair, which dangles from his rapacious hand, while at the table, the distraught Belinda is being consoled by Thalestris. Beneath the table the vague lumpy form is perhaps the faithful Shock at his mistress's feet. The print, in spite of its miniscule size, outlines the scene efficiently. It is a contemporary interpretation that sees the poem in bluntly literal terms as an episode in London's beau-monde, with nothing of the fantasy and wit of Pope's imagination.

After the initial publication of the poem (1714—23) although no separate edition was published again until the end of the century, it continued to be included in the frequent reprintings of Pope's works.

1. Pope, *Works*, ed. J. Warton (1797), I, 317.
2. Ronald Paulson, *Hogarth: His Life, Art, and Times* (New Haven and London, 1971), I, 69. The print, identified by Horace Walpole, is now in the Lewis Walpole Collection.

10. [William Hogarth?], The Rape of the Lock [1717?], engraving

The nine-volume edition in 1751, prepared by his literary executor, William Warburton, embellished the poem with a single plate designed and engraved by Anthony Walker. [*Plate 11*] As both artist and engraver, Walker was well known for his small book-illustrations, and often—as in this plate—used more etching than line engraving. Although this enabled him to achieve a greater delicacy his plates lack sharp precision and are sometimes indistinct.[3]

What a tame and decorous tea-party this artist has drawn! The Baron, standing on tip-toe, prepares to snip off Belinda's irresistibly tempting lock of hair at an intimate, humdrum—one can hardly resist saying tepid—tea party. The draughtsmanship of Walker, a Yorkshireman, has a French elegance that conveys nothing of the mock heroics flaunted in the ungainly 1714 plate of the same scene, though a note of excitement comes from the horrified maidservant in the background as she sees the Baron about to cut the lock. A different kind of excitement is taking place above the mantle, where a pair of large cupids frame an oval painting of a satyr entertaining

3. Hammelmann, *Book Illustrators in Eighteenth-Century England*, p. 97.

11. Anthony Walker, Pope's *Works* (1751), Vol. I, engraving

with music two indistinct female figures. In contrasting this suggestive pagan scene with the decorum of the drama below, the artist extracted an important theme of the poem hidden beneath its urbane social surface.

In the next collected edition of Pope's works, in 1757, the poem was illustrated by two plates. Although unsigned they can firmly be attributed, on stylistic and documentary evidence, to Samuel Wale, the prolific illustrator,[4] who invariably let his drawings be engraved by a professional craftsman onto copperplate.[5]

In the first illustration [*Plate 12*], the climactic scene of the poem, Sir Plume—recognized by his cane and snuffbox—faces the exultant Baron. But the heroine's personality and attitude have undergone a sharp stylistic change. In Du Guernier's 1714 design for that scene Belinda emotes with theatrical rhetoric; she is a robust tragic figure, reflecting Pope's description of eyes flashing lightning and screams of horror. In Wale's drawing she is a drooping, sentimental heroine, holding an oversized handkerchief to catch her copious tears, while behind her stands a friend (Thalestris, no doubt) prepared to administer smelling-salts. Belinda has been overtaken by the movement known as sentimentalism, which since its origin in the late seventeenth century had developed rapidly and was fully exploited in drama (such as Steele's comedies) and fiction (such as Richardson's *Pamela* and *Clarissa*). Belinda has thus been converted from a frivolous, mocked belle to a sentimental, lachrymose heroine-victim.

In Wale's drawing of the Cave of Spleen [*Plates 13, 13a*] the Queen—whose face is remarkably like Belinda's in the other plate—'sighs on her pensive Bed'. At her left are the two named Handmaids: Affectation, facing a dressing-table mirror, and Ill-nature, clutching lampoons to her bosom. But instead of interpreting Pain and Megrim as afflictions suffered by the Queen, the illustrator has personified them—at the Queen's right—as two additional handmaids or lesser attendants. He also departed from the 1714 illustration of the line 'Maids turn'd Bottels, call aloud for Corks'. For where that plate had shown the maid-bottles corked with men's heads—suggesting that the call had been answered—the bottles are now more innocently surmounted by the heads of the maids themselves.

4. The original drawing of 'The Cave of Spleen' is accompanied by a label in a distinctive hand ascribing it to Samuel Wale. Annotations in the same hand and dated 1863 accompany the collection of Wale's drawings (in the British Museum since 1859) for the illustrations in the *Tyburn Chronicle* and *Newgate Calendar*.

5. Hammelmann, p. 89.

12. [Samuel Wale], Pope's *Works* (1757), Vol. I, engraving

13. [Samuel Wale], The Cave of Spleen (1757), drawing

14. John Hamilton Mortimer and Charles Grignion, frontispiece to Pope's *Poetical Works* (1776), engraving

13a. [Samuel Wale], Pope's *Works* (1757), Vol. I, engraving

Here the illustrator's interpretation of the text may perhaps be taken as a sign of the change in moral climate. Pope's sexual wit, a legacy from the moral libertinism of Restoration drama and poetry, could be evaded in this way by an illustrator serving middle-class readers of more stringent morality.

The other objects on the floor of the Cave of Spleen presented less difficulty to the illustrator: in the foreground the oversized, living tea-pot and the talking goose-pie, while Umbriel on the right takes leave of the Queen. Although the poem confines this scene to the grotto-like cave, the illustrator adds (in an upper corner) an architectural element resembling a Wren church steeple rather than Hampton Court Palace, Umbriel's destination, where the next scene will take place.

In the later eighteenth century, as new currents of thought and sensibility were preparing the way for the romanticism that flowered at the turn of the century, Pope's reputation as a poet declined. One of the pioneers in this reassessment was Joseph Warton, who in 1756 published the first volume of his *Essay on the Genius and Writings of Mr. Pope*. After comparing Pope to Spenser, Shakespeare, and Milton, he confidently demoted him to a lesser species of poet, though he singled out *The Rape of the Lock* as exceptional: 'in this composition, POPE principally appears a POET; in which he has displayed more imagination than in all his other works taken together'.[6] Warton's low opinion of Pope's rank may have been premature, for he found so little popular support that he waited twenty-six years before publishing the second volume of his *Essay*.[7] Such disparagement of Pope was gently satirized by Samuel Johnson when (in 1759) he has this to say of Dick Minim, a trendy critic: 'Pope he was inclined to degrade from a poet to a versifier, and thought his numbers rather luscious than sweet'.[8]

The universal popularity and appeal of the poem was unequivocally proclaimed by Johnson himself in his 'Life of Pope' (1781), where he wrote: 'To the praises which have been accumulated on *The Rape of the Lock* by readers of every class, from the critick to the

6. *Essay on the Genius and Writings of Mr. Pope* (1756, 1782), I, 248.
7. This explanation for Warton's delay, as given by Samuel Johnson in 1772 (reported by Boswell), has recently been challenged (Joan Pittock, 'Joseph Warton and his Second Volume of the *Essay on Pope*', *Review of English Studies*, N.S., XVIII (1967), 268, 272–3).
8. *The Idler* No. 60 (The Yale Edition of the Works of Samuel Johnson, Vol. II, 1963, p. 187).

waiting-maid, it is difficult to make any addition'.[9] Other critics, more attuned to the new romanticism, praised only the wit of the poem rather than its qualities of imagination and invention. That wit, declared Hugh Blair (in 1783), rendered the poem 'the greatest master-piece that perhaps was ever composed, in the gay and sprightly Style. . . .'[10] This characterization of the poem remained the most persistent interpretation for most artists at the end of the eighteenth century.

The publication of literary illustration was greatly stimulated in the last quarter of the century by John Bell, publisher, bookseller, printer, typefounder and journalist, whose inexpensive editions of the English classics were widely distributed. Besides designing the typography of his books with the greatest care, he embellished the little 16mo volumes with frontispieces designed by the leading illustrators of the day. Between 1777 and 1792 he published *The British Poets from Chaucer to Churchill* in 109 volumes.[11] In the former year, he issued Pope's poems, with a frontispiece by John Hamilton Mortimer. [*Plate 14*] The artist endowed his heroine with the luxurious sensuality of a courtesan, her bare bosom displayed, as she placates her lap-dog. This alteration of Belinda from a starchy, frivolous court belle may not be what the poem's text ordinarily brings to mind; but to a painter accustomed, as Mortimer was, to historical and romantic subjects, an opulent, Venus-like figure came more easily to hand. The omission of 'busy *Sylphs*' who perform the labours of dressing Belinda removes it further from a literal representation of Pope's text.

The transformation of Pope's coquette into a sensual temptress (derived from Renaissance models, no doubt) attracted other illustrators at this time. The Rev. Matthew Peters, who was first a painter and member of the Royal Academy and later a clergyman, is best known for the sentimentalized scenes he contributed to Boydell's Shakespeare Gallery. He painted a portrait which he entitled 'Belinda' and for which a Miss Bampfylde sat; it was published as an engraving in 1777. [*Plate 15*] Of all the portraits of Pope's heroine this is the most lubricious.

A further example of Belinda as sensual temptress was published

9. *Lives of the English Poets*, ed. G. B. Hill (Oxford, 1905), III, 232. In his reference to 'waiting-maid' Johnson was perhaps echoing the tribute to the poem in John Gay's *Trivia* (1716, p. 49): 'Pleas'd Sempstresses the *Lock*'s fam'd *Rape* unfold'.

10. *Lectures on Rhetoric and Belles Lettres* (1783), II, 369.

11. Stanley Morison, *John Bell, 1745–1831* (1930), pp. 15, 96.

I. Thomas Stothard, Belinda on her Voyage to Hampton Court (1798), watercolour

II. Thomas Stothard, Belinda on her Voyage to Hampton Court (1804), watercolour

III. Thomas Stothard, The Rape of the Lock (1798), watercolour

IV. Thomas Stothard, The Battle for the Lock (1798), watercolour

15. Matthew William Peters and R. Dunkarton, 'Belinda (Miss Bampfylde)' (1777), engraving

in the 1796 edition of Bell's British Poets, where the title page of Pope's poems is embellished by a drawing of Belinda asleep. [*Plate 16*] Above the encircled scene, Cupid—symbolically crowned by a pair of scissors and flanked by a cross-bow and quiver of arrows—presides like a deity over the bare-breasted Belinda enjoying her delicious dream, and smiling with a Greuze-like expression of ecstasy: sensuality and sentimentalism combine. At one side Shock sleeps luxuriously on a fringed pillow, while Ariel hovers above, sustained by gauzy wings as he points his wand to orchestrate Belinda's fateful dream. Through the open-curtained bed can be seen the toilet-table, the next scene of the poem. The artist's plate, modestly signed with only the name Graham, was by John Graham, a Scottish historical painter active in London from 1780 to 1797.

In the latter year Graham exhibited a painting in the Royal Academy labeled 'From Pope's Rape of the Lock', which was

16. John Graham, frontispiece to Pope's *Poetical Works* (1796), engraving

17. William Hamilton, The Graces and a Satyr Train (1798), drawing

probably the source of the engraved plate. This painting and its derivative engraving point up a newly developed practice in book illustration. Earlier book illustrators had drawn their designs only for that purpose, regarding them only as preliminaries for the plates; hence they were usually the same small size and executed in black and white, often with grey wash, to make easier the engraver's work of transferring the design to the copper plate.

But following the establishment of the Royal Academy in 1768, with its annual shows, artists became aware of the practical and artistic advantages of painting for exhibition large, brilliantly coloured oil or watercolour literary illustrations; and after the engraver had copied them (in greatly reduced size, of course) the original pictures could be sold. Such painting thus became more prestigious and more profitable, and induced ambitious, successful artists to turn their talents to illustration because it did not confine them to the limited and poorly paid work for book publishers alone.

While wealthy collectors could hang the original paintings on their walls, middling and humble folk could buy for a very modest price engravings of the same pictures. It was Hogarth who initiated this wave of popular art with his series of engravings *The Harlot's Progress* (1733). In later decades the trade in selling separate engravings grew to awesome proportions. By the time of John Boydell (in the 1780s) the export of historical prints alone brought to England in one year the incredible sum of £200,000.[12] Literary illustrations, apart from being sold as prints, could be further reduced in size to be included as book illustrations with the published literary work. The outstanding example of this pattern and practise was Alderman Boydell's Shakespeare Gallery, opened in 1789. In his three-pronged enterprise he commissioned large paintings from all the leading artists of the time; and after the canvases were exhibited in his Gallery on Pall Mall, they were engraved to be sold as a collection of prints, and finally (along with newly commissioned designs) were engraved in smaller format to serve as illustrations for his 1802 edition of Shakespeare.

Artists, while painting for picture collectors and enterprising publishers, were affected in varying degrees by the artistic and intellectual currents of their time. How can this be assessed? One guide, itself a useful sample of current thinking at the end of the eighteenth century, was suggested (in 1807) by the engraver John

12. John Pye, *Patronage of British Art* (1845), pp. 42–4, 244.

Landseer, father of the popular Victorian painter. The artist, he wrote, can accept 'the prevailing notions of Beauty and Propriety, which exist in the minds of his contemporaries, and endeavour gradually to refine and raise them to the level of his own'; or, 'warmed by a nobler enthusiasm', he can exercise his profession for posterity, and strive for posthumous glory.[13] The artists who illustrated *The Rape of the Lock* on the eve of the nineteenth century can be separated into these two groups. A talented craftsman like Thomas Stothard was satisfied to reflect the dominant taste of his time, while a genius like Henry Fuseli, by consulting his own personal taste, expressed a more striking, timeless interpretation.

The year 1798 is an important milestone in the history of English poetry. The *Lyrical Ballads* of Wordsworth and Coleridge contained, according to its Advertisement, 'a natural delineation of human passions, human characters, and human incidents'. More pointedly, in the preface to the 1800 edition Wordsworth defined the incidents and language as arising from common life. By no stretch of criticism or imagination could *The Rape of the Lock* be set under such a rubric. But Wordsworth's creed was far from being universal; and *The Rape of the Lock* still ranked as a masterpiece of some sort, if not of Wordsworth's sort. In that same year (1798) a London publisher believed it important enough to be issued in an illustrated edition, the first separate one since its initial publication in 1714.

Francis Isaac Du Roveray, born in 1772, was the son of a Swiss from Geneva settled in England,[14] and became a very active publisher as a young man still in his twenties. Between 1798 and 1804 he published, besides *The Rape of the Lock*, editions of Thomas Gray's *Poems, Paradise Lost*, James Thomson's *The Seasons*, and Pope's *Poetical Works*. All of them were lavishly illustrated by a group of artists, whom he also commissioned to design the plates for Pope's mock-epic.

The title page of his 1798 edition reads: *The Rape of the Lock, an Heroi-Comical Poem. By A. Pope. Adorned with Plates*. Besides the frontispiece, five full-page plates 'adorned' the slim octavo volume, one for each canto.[15] In the long preliminary 'Advertisement', which

13. *Lectures on Engraving* (1807), pp. 219–20; quoted in Ralph Cohen, *Art of Discrimination* (Berkeley, 1964), p. 271.

14. He died in London in 1849, aged 77 (*Annual Register*). A letter in French from his father, dated London, 1783, discusses the settlement of *émigrés* from Geneva (BL Add. MS 33100, f. 283).

15. When showing his edition of *The Rape of the Lock* to Joseph Farington, he said that he

he probably wrote himself, Du Roveray outlined the history of the poem, citing various critics and quoting their opinions. In his own 'eulogium' he stated that 'no work contains such delicate, and, at the same time, such forcible strokes of wit, free from coarseness and ribaldry, which are too often mistaken for wit . . .' (pp. xv–xvi). Yet he did not remove from the text the smutty couplet that Pope's contemporaries regarded as coarse; nor in fact did any of the later editors, including a Victorian editor who prepared a text for young people.[16] Perhaps all these editors simply averted their attention from the implications of the couplet, for deleting it would have drawn attention to it.

The Advertisement was followed by Thomas Parnell's verse tribute 'To Mr. Pope' (1722), to add some bulk to the thin volume. The artist who supplied the frontispiece [*Plates 17, 17a*], William Hamilton, chose to illustrate not a scene from the poem but rather the couplet from Parnell that was printed beneath the plate:

The Graces stand in sight; a Satyr train
Peeps o'er their head, and laughs behind the scene.

His hackneyed design of the three Graces displays them in coy poses that effectively hide their pudenda, while three satyrs spy on them lustfully. Since the Graces, daughters of Zeus and associated with Apollo as well as Aphrodite, are traditionally chaste and inviolate, the satyrs are mere voyeurs.

A painter, almost certainly William Hamilton, was more forcibly inspired by Pope's poem to sketch a small oil painting, which has never been engraved or otherwise reproduced. [*Frontispiece*] The Baron, in the contemporary dress of a Regency rake, exults in his moment of triumph, the ravished lock of hair in one hand, the scissors in the other, while Belinda hysterically reaches out in vain to retrieve the prize. The whole composition vibrates with a febrile intensity more characteristic of the mannerist distortion of Fuseli, whom Hamilton imitated in his later years (he died in 1801).

The artist who illustrated Canto I (in the 1798 edition) was Edward Francis Burney. [*Plate 18*] Here is Belinda at the completion of her toilette, when 'busy sylphs surround their darling care', arranging her hair and gown. In the poem she reads the *billet-doux* in bed as soon as she opens her eyes; now she is perhaps rereading it,

had five hundred subscribers to it, which would clear his expenses (16 Dec. 1798, typescript of Farington diary in British Museum, p. 1394). Using the same illustrations, he published another edition in 1801.

16. ed. W. C. Macready (1849).

17a. William Hamilton and Francesco Bartolozzi, frontispiece to *The Rape of the Lock* (1798), engraving

18. Edward Francis Burney and Francesco Bartolozzi, Canto I of *The Rape of the Lock* (1798), engraving

while she soothes the jealous Shock. Burney had designed a very similar picture as the frontispiece for Cooke's Pocket Edition of Pope's *Works*, published in 1795. It is curious that he should have designed a close variant and that Du Roveray should have accepted it for his pretentious 1798 edition.

The plates for Cantos II, III, and V were designed by Thomas Stothard, one of the most prolific illustrators of the time. Besides his many illustrations for Pope's works, noted by his biographers, Stothard exhibited at the Royal Academy in 1806 a composition entitled Belinda, and at the British Institution in 1841 five sketches based on the poem.[17] These had probably been used for the published illustrations. The small size of Stothard's watercolours as well as of the Hamilton drawing was undoubtedly suggested by Du Roveray, who had specified it for Fuseli's illustration (see below, p. 45). Since the engraved plates for the edition were to be exactly this size the engraver could easily transfer the design onto copper more accurately and rapidly.

For Canto II Stothard painted a watercolour of the utmost delicacy [*Plates I, 19*], sketching the merest suggestion of the ship's mast and rigging to set the scene on the 'painted Vessel'. Ariel, with purple pinions and azure wand, is at his commanding post. This sylph is clearly a folk-lorist fairy.[18] Previous illustrators, in trying to visualize the invisible sylphs, had recalled comparable creatures from ancient and Renaissance art: cupids, cherubim, and putti. Stothard, in converting them to winged fairy-like creatures, is one of the first illustrators to furnish fairies with wings like those of a butterfly, an innovation followed by later artists.[19] An anecdote relates that early in his career Stothard was puzzled as how best to represent such a fanciful creature as a sylph. A friend suggested he give the sylph a butterfly's wing. ' "That I will," exclaimed Stothard; "and to be correct, I will paint the wing from the butterfly itself." He immediately sallied forth, extended his walk to the fields some miles distant, and caught one of those beautiful insects: it was of the class called the peacock.' This was the beginning of his interest in butterflies and their infinitely varied and miraculous colouring.[20]

17. Mrs. [Anna Eliza] Bray, *Life of Thomas Stothard* (1851), pp. 226, 243, 236.
18. K. M. Briggs, *The Fairies in English Tradition and Literature* (Chicago, 1967), p. 156.
19. Briggs, pp. 157, 161.
20. Bray, pp. 31–2. A scholar has recently suggested that Blake 'may have been' the friend who advised Stothard, and that Blake himself was 'perhaps influenced by Fuseli, an entomologist since age 12' (John Adlard, *The Sports of Cruelty, Fairies, Folk-Songs, Charms and Other Country Matters in the Work of William Blake*, 1972, pp. 77–8).

19. Thomas Stothard and Anker Smith, Canto II of *The Rape of the Lock* (1798), engraving

20. Thomas Stothard and William Bromley, Canto II of 'The Rape of the Lock' (1804), engraving

His imagination was less challenged by the humans in the poem. Belinda, the sparkling cross visible on her breast, is attended by the Baron, who, in the poem, is not mentioned as accompanying her. Stothard later painted a different version of this scene for the 1804 edition of Pope's works, also issued by Du Roveray. [*Plates II, 20*] In this he removes Belinda from her circle of friends to elevate her on a red carpet as a queen-like presence receiving deferential homage from her fellow mortals and vigilant attention from the sylphs. Pope's intention in writing the poem, as he declares in his dedicatory epistle, was 'only to divert a few young Ladies, who have good Sense and good Humour enough, to laugh not only at their Sex's little unguarded Follies, but at their own'. In the mid-century illustrations, as we have seen, sentimentalism had begun to alter the contours of Belinda's portrait. Stothard's illustration transforms her differently: from the mocked heroine of a mock-epic into a genuine heroine whose maidenly dignity is grossly violated by the beastly Baron. This transformation becomes more marked in subsequent nineteenth-century interpretations.

In his design for Canto III in the 1798 edition [*Plates III, 21*] Stothard chose the difficult, challenging moment when the Baron is about to snip off the lock and a 'wretched sylph too fondly interpos'd' (but after being cut in twain was soon united again). Stothard emphasizes the valiant attempt by the sylphs to protect Belinda, while Ariel—visible in the upper right—retreats in amazement and confusion as his power expires. In the poem Ariel guards Shock; Stothard departs from the text by banishing Shock from the scene and placing Ariel on top of the tea-table.

In illustrating Canto V—the battle scene—Stothard more drastically departs from the text. [*Plates IV, 22*] For in the back, on the right, the furious Belinda is subduing a man by throwing a charge of snuff to his nose. Yet the Beau kneeling at the right resembles the Baron in the other drawings. Here he seems to be cast as Dapperwit or Sir Fopling, about to expire (metaphorically) at the feet of Thalestris. Stothard has also altered the character of the gnome Umbriel, visible on the upper border. In the poem he triumphantly 'Clapp'd his glad wings' as he views the battle; here he looks distinctly worried and unhappy. In this way Stothard has again softened the satiric tone of the poem into a sentimental one: while Pope's mischief-making gnome rejoices to see the vain belles and beaux unhappy and frustrated, Stothard's gnome has been won over by Belinda's distress and shares it.

21. Thomas Stothard and James Neagle, Canto III of *The Rape of the Lock* (1798), engraving

22. Thomas Stothard and William Bromley, Canto V of *The Rape of the Lock* (1798), engraving

In the reviews after the book's publication one critic praised the 'beautiful' plates by 'artists of the first class'.[21] In two other reviews Stothard was the only artist singled out (perhaps because he had provided three illustrations, the others only one each). *The British Critic*, while conceding that the volume was 'eminently beautiful', accused Stothard of having mistaken the character of the sylphs by delineating them as 'fine and full-grown Cupids'.[22] The *Gentleman's Magazine*, although it repeated the criticism that he had 'mistaken Sylphs for full-grown Cupids', at least conceded that 'perhaps none could express such aërial beings, if expressible, as Fuseli'.[23] Yet ironically, Fuseli designed the plate for Canto IV, 'The Cave of Spleen', a scene where no sylphs appear. His design for this plate stands out in conspicuous contrast to the others as the work of a master illustrator. It is also remarkable because by chance there has survived the full record of how it was commissioned by Du Roveray and how Fuseli went about executing it.

21. *Monthly Review*, 29 (1799), 101.
22. 13 (1799), 310.
23. 69, i (1799), 417.

IV

Fuseli and *The Rape of the Lock*

Henry Fuseli was no mere illustrator, though he plied that trade indefatigably, but one of the great European artists of his time. His interest in many literatures, not only English, gives his œuvre a broad range beyond his English contemporaries whom he had joined as a young Swiss émigré. Sacheverel Sitwell calls him a 'strange character, who poised between literature and painting, who was more interested to write of painting, and paint when under the inspiration of poetry'.[1] The literary works that generally inspired his imagination were the classics of *Weltliteratur*—the *Nibelungenlied*, the works of Homer, Virgil, Ovid, Dante, Spenser, Milton, and above all Shakespeare. He illustrated these on a heroic scale in paintings for exhibition; single-handed he painted forty-seven canvases for his Milton Gallery and contributed nine to Alderman Boydell's Shakespeare Gallery. He designed illustrations as well for lesser English writers—Smollett's *Peregrine Pickle* (1769), Erasmus Darwin's *The Botanic Garden* (1789–91), Cowper's *Poems* (1806); and for Du Roveray, besides *The Rape of the Lock*, Gray's *Poems* (1800), *Paradise Lost* (1802), and Thomson's *Seasons* (1802).

With typical Romantic taste Fuseli had no high opinion of Pope's poetry—'metrical and rimed prose', he called it.[2] On another occasion he disparaged Pope as one who counted 'his verses by thousands, who has not learnt to distinguish the harmony of two lines from that of a period—whom dull monotony of ear condemns to the drowsy psalmody of one returning couplet'.[3] But he regarded *The Rape of the Lock* as exceptional, the only poem in which Pope showed 'poetic genius', and for which Pope deserved the honored epithet: 'A Poet is an inventor; and what has Pope invented, except the Sylphs?'[4]

1. *Narrative Pictures A Survey of English Genre and Its Painters* (1937), p. 58.
2. Eudo C. Mason, *The Mind of Henry Fuseli Selections from His Writings* (1951), p. 114.
3. John Knowles, *The Life and Writings of Henry Fuseli* (1831), III, 104.
4. Knowles, I, 358. Yet Fuseli imitated Pope's *Dunciad* in four fragments, composed in the 1780s, and included some echoes of *The Rape of the Lock* (Eudo C. Mason, *Unveröffentlichte Gedichte con Johann Heinrich Füssli*, Zürich, 1951, pp. 57–8).

Whether or not Du Roveray knew that Fuseli held this grudging opinion of Pope does not matter—although if he did, would he not have asked him to illustrate a scene in which sylphs appear? Through his correspondence with Fuseli about that artist's commission we can follow in detail the progress of their negotiations.

On 1 March 1798 Du Roveray wrote to Fuseli (from Great Saint Helen's in the City): 'I beg the favor of your informing me whether you would do a Drawing for me, in water colours, *4 inches by 3 upright,* from Pope's Rape of the Lock representing *the Cave of Spleen, with Umbriel receiving from the goddess the bag & vial*; likewise what you would charge to make a *finish'd* drawing of it; and what time the same would require? If you objected to make so small a drawing I should be obliged to you to mention the size that would be agreeable to you.'[5] Du Roveray may have chosen 'The Cave of Spleen' for Fuseli because he regarded its macabre and grotesque subject more congenial to the artist's style than the delicate and benign sylphs. And as a practical requirement he wanted the drawing to have exactly the same dimensions as the plate to be engraved for the printed book.

Fuseli accepted the commission, and availed himself also of Du Roveray's liberality in letting him choose a different size if he wished. He replied promptly (on 3 March) and to the point: 'I have had the honour of your Note & have considered the Subject you propose. It is a good one, and I could not promise to do it much justice in an upright Drawing of less than 16 inches by 14 or 15. The price ten Guineas; the time a Month.'[6] He probably preferred a larger size—four times what Du Roveray had requested—for two reasons: he could draw more and finer details than in a small drawing, and he could exhibit it (as indeed he did) after the engraver had copied it.

Du Roveray proved to be a truly liberal patron, replying on 5 March: '. . . I beg you will execute for me the drawing mention'd in my former letter; only I could wish the dimensions not to be quite so square as those you propose, say *16 inches by 12 upright* tho' if you could make a better drawing of it 16 by 14 you have my consent so to do. I am glad to see that you are pleased with the subject, and am persuaded you will render it justice.' He concluded with an invitation to Fuseli to call on him so that he could show him some drawings and prints.[7]

5. Copy by Du Roveray, MS in Greater London Record Office.
6. MS in Pierpont Morgan Library.
7. Copy by Du Roveray, MS in Greater London Record Office.

By the end of the month Fuseli had not got very far with his commissioned drawing, and on 31 March wrote to Du Roveray to explain why: 'The necessity of finishing a picture for the exhibition at Somerset house [annual exhibition of the Royal Academy][8] has not permitted me to go beyond a kind of mental composition, and something of a slight sketch in the execution of Your Drawing. I certainly should prefer Oil colour on *Strained paper* as uniting with all the advantages of Water colours those of warmth & durability: but before I can mention a price, I wish to be informed whether you mean to adhere to the *Size* agreed on for the drawing, or can allow something more; which without much increasing the expence, would add greatly to the elegance of the whole?'[9]

This letter evidently crossed one from Du Roveray's in the post, for Fuseli wrote again on 2 April (thanking him for his letter of 31 March): 'In reply to your favour of Saturday I must observe that the increase of Size you propose will be of great Service to render the finish of the Picture more "exquisite",—at least in my hands; and in that Case you will not think thirty Guineas a high price. In a work of Taste Time must or ought to wait on Fancy, but a month or six weeks might probably answer the purpose.'[10] Again Du Roveray proved to be generous and tolerant. 'In answer to your favour of yesterday,' he wrote on 3 April, 'I think thirty Guineas a very high price for so small a picture; however, as you are the best judge of the work, and I wish to have something as exquisitely done as possible, I shall not object to it, persuaded that the great encrease of price (for you know I first reckoned upon having a finished drawing for 10 Gs) will be fully compensated by the execution. When you have pleased yourself as to the composition I shall be glad to see a sketch of it.'[11] In reply Fuseli was willing to 'increase the Picture to 24 inches or upwards, upright, with a proportionate width. I shall do it with pleasure & without any additional demand', he conceded.[12]

8. No. 220 in exhibition of 1798: 'Richard III in his tent the night preceding the battle of Bosworth. . . .'
9. MS in Hilles Collection, Beinecke Library, Yale University.
10. MS in Princeton University Library.
11. MS in Duycinck Collection, New York Public Library. In offering Fuseli ten guineas (let alone agreeing to thirty) Du Roveray was being generous. For at about the same time, John Bell, the publisher of *British Poets*, asked Stothard to submit two or three designs, from which Bell selected one and paid one guinea for it (John Pye, *Patronage of British Art*, 1845, p. 247). Perhaps Du Roveray was known as a generous patron, for Stothard had asked two guineas for a design in India ink and three for one in colour (2 Feb. [1797], MS in Princeton University Library).
12. MS in National Library of Scotland.

Fuseli was evidently the only one of the four artists engaged in this enterprise who chose not to follow Du Roveray's initial request for a drawing the same size that the engraved plate would be. The drawing made by Hamilton for the frontispiece and the three watercolours by Stothard conform to Du Roveray's suggested dimensions. At least Fuseli's differences with Du Roveray were amicably settled, and he set to work in earnest. A fortnight later (on 16 April) he invited Du Roveray to call on him in Queen Anne Street to see the progress he had made in his picture.[13]

By the middle of the summer, when the book was in production, Du Roveray had a different problem to face. His printer Thomas Bensley wrote to him (on 18 July): 'Upon casting off the Rape of the Lock, I find it will only make $4\frac{1}{2}$ Sh[ee]ts even allowing a Bastard Title to each Canto. What No. of pages the Preface & the Life you spoke of may contain I know not, but sh[oul]d think scarcely so much as a Sheet & $\frac{1}{2}$— Shall therefore not order the paper until I hear from or see you again.'[14] Du Roveray had already prefixed the poem with Parnell's verse (seven pages) as well as Pope's dedicatory letter to Mrs. Arabella Fermor (Belinda); yet the book still seemed to the printer to be too slight.

Instead of 'the Preface & the Life', Du Roveray prepared the 'Advertisement'—ten pages in length—most of it a critique of *The Rape of the Lock* (with nothing of Pope's life). It contained a long extract from Samuel Johnson's criticism of the poem (in his 'Life of Pope'): 'The heathen deities can no longer gain attention: we should have turned away from a contest between Venus and Diana. The employment of allegorical persons always excites conviction of its own absurdity; they may produce effects, but cannot conduct actions: when the phantom is put in motion it dissolves: thus Discord [in Boileau's mock-epic *Le Lutrin*] may raise a mutiny; but Discord cannot conduct a march, nor besiege a town.' Before having it printed Du Roveray sent the 'Advertisement' to Fuseli for his opinion of it.

In his reply, on 20 July, Fuseli first thanked Du Roveray for the preface, 'written with equal propriety & ease' and then wrote that he could only disagree with 'the quotation from Dr Johnson, the insertion of his shallow remark on the admission of Allegorical

13. MS in Princeton University Library.
14. MS in Edinburgh University Library.

Actors; Spleen & her attendants being nothing else'.[15] Du Roveray took Fuseli's stricture seriously enough to print it as a footnote to his text: 'This remark of Dr. Johnson's seems rather shallow, and it is certainly ill applied; for what are *Spleen* and her attendants but allegorical actors?' In this way Fuseli shared the editing of the 'Advertisement'.

Du Roveray may have faced further delays in publishing the little volume, for (with 1798 on its title page) it was not issued until January 1799.[16] And in the spring of that year the exhibition catalogue of the Royal Academy, after listing as No. 157 Fuseli's 'The Cave of Spleen, for Mr. F. J. Du Roveray's new edition of Pope's Rape of the Lock', printed ten lines from the poem, beginning 'The goddess with a discontented air'.

As engraved by Thomas Holloway [*Plate 23*] the illustration conveys an impression of cool neo-classicism, a contrast to the fussy, almost rococo style of the other illustrations in the volume.[17] The design centers on the goddess Spleen, as she dismisses the gnome Umbriel. Her pose and gesture are similar to those of Queen Katherine in Fuseli's illustration for Shakespeare's *King Henry VIII* (IV.ii), where the unhappy queen, flanked by two handmaids, holds up her hands to heaven as her dream-vision vanishes.[18]

Spleen's two authentic handmaids are clearly shown: Ill-nature, turning her back to the Queen, carries a prayer-book in her hand, Fuseli's neat interpretation of Pope's 'store of pray'rs for mornings, nights, and noons'. Affectation, kneeling on the other side, exhibits appropriate mannerisms, one hand holding her fan and the other fondling her lap-dog, that important adjunct to female vanity. This addition by Fuseli links Affectation to Belinda, whose lap-dog is her most precious possession. Fuseli has also personified Pain (above Spleen's right hand) and Megrim (directly above her head): a misreading sanctioned by the previous illustrator (Samuel Wale). Of the 'unnumber'd throngs' in the Cave of Spleen, Fuseli drew a few: a snake on rolling spires, a goose[berry] pie that (presumably) talks,

15. MS in Royal Academy Library, Jupp *Catalogue of Exhibitions*, III, f. 19.
16. *Monthly Epitome and Catalogue of New Publications*, III, 34.
17. The London publishers Cadell and Davies were so impressed by the illustration that they wrote to William Roscoe, Fuseli's friend and patron, on 3 Jan. 1799: 'A very admirable Proof of what may be done on a small Scale from a Picture of Mr Fuseli's, has lately appeared in a new Edition of Pope's Rape of the Lock' (Roscoe Papers 627, Liverpool Public Libraries).
18. Gert Schiff, *Johann Heinrich Füssli 1741–1825* (Zürich and München, 1973), No. 729 (painted in the 1780s). Hereafter Schiff.

V. Henry Fuseli, The Dream of Belinda [1780–90], oil

23. Henry Fuseli and Thomas Holloway, Canto IV of *The Rape of the Lock* (1798), engraving

and next to it a living teapot. It is surprising that, with his vigorous erotic imagination, he neglected to draw the maids turned bottles calling aloud for corks. He may have been restrained by Du Roveray's scruples, who in the 'Advertisement' praises the poem as 'free from coarseness and ribaldry'. Fuseli's illustration also contains creatures from his own imagination and graphic vocabulary, rather than from Pope's poem: a miniature female troubadour—a motif he used in a Shakespeare illustration[19]—who serenades Affectation, and the figure of a man, on the far left, silhouetted against a glowing light.[20]

Since this plate 'adorned' only Canto IV of the poem its subject was necessarily confined to that section. But in the 1780s Fuseli had used the entire poem as the subject of a large oil-painting. [*Plate V*] This brilliant work is neither illustration nor adornment but rather an improvisation on themes in the poem mingled with symbols from Fuseli's own world of dreams and fantasy. The difficulties of interpreting this painting are only too evident, for until a few years ago it was known as 'Queen Mab'; an art historian then labeled it (correctly, I believe) 'The Dream of Belinda';[21] the leading Fuseli scholar today entitles it 'Belinda's Awakening' (which is not the same), while another Fuseli scholar mistakenly assumes it to be a variant of 'The Cave of Spleen'.[22]

'One of the most unexplored regions of art are dreams' is one of Fuseli's aphorisms;[23] he could have been thinking of the dream he conjured up for the heroine of Pope's poem. His painting, its subject matter from Pope's poem and his own fantasies, also resonates with echoes of Shakespeare, his most pervasive and obsessive literary influence. The main, recumbent figure is certainly Belinda asleep (her eyes firmly shut) as her 'morning-dream . . . hover'd o'er her head'. The sparkling cross she will wear on her voyage to Hampton Court is already suspended from her neck. Her distinctive posture is similar to that of Michelangelo's 'Night' on the Medici tomb, which Fuseli

19. A Midsummer Night's Dream (Schiff No. 885).
20. The plate is described in Jean H. Hagstrum, *William Blake Poet and Painter* (Chicago, 1964), p. 70.
21. Harold D. Kalman, 'Füssli, Pope and the Nightmare', *Pantheon*, 29 (1971), 226–34. Kalman dates its composition as 1782, Schiff (No. 751) as between 1780 and 1790. In my discussion of the painting I accept some of their interpretations but differ in others.
22. Schiff, I/I, p. 632; Peter Tomory, *The Life and Art of Henry Fuseli* (1972), p. 112.
23. John Knowles, *The Life and Writings of Henry Fuseli* (1831), III, aphorism 231.

had sketched in 1777.[24] It is even closer to Fuseli's own drawing 'Sleeping Woman and Cupid', dating from the same decade as the painting.[25]

In that drawing the Cupid in the air aiming his arrow at the sleeping woman is equivalent to the figure suspended in the air above Belinda: the human-sized, androgynous Ariel. Pope had borrowed the name from *The Tempest*, in which Ariel has a comparable function as Prospero's attendant spirit; and Fuseli, who needed no prompting to pay tribute to Shakespeare, suspends Pope's Ariel in the same position (though without wings) as in several of his previous versions of Shakespeare's Ariel, the earliest in 1774.[26]

On the dressing table that Ariel holds before the sleeping Belinda a lover's letter is pushed forward, projecting over the edge: thus, as in Pope's lines, Belinda's eyes will first open on a *billet-doux*. A mirror and cosmetic containers complete the furnishings of the table. For further evidence of Belinda's vanity, Fuseli placed on the far right of the painting—and hardly visible—a casket such as is used for holding jewels, though on it is perched a glaring owl.

The large, dun-coloured male figure who plunges toward the viewer is the gnome Umbriel carrying (as the poem tells) 'a branch of healing spleenwort in his [right] hand'. Fuseli reinforced the allusion to spleenwort by strewing in Umbriel's hair the characteristically arrow-shaped leaves of this fern-like plant (genus *Asplenium*). Again, a parallel to Shakespeare is immediately apparent: the mischievous Puck (or Robin Goodfellow) of *A Midsummer Night's Dream*, painted by Fuseli for the Boydell Gallery,[27] is similar in body and facial expression to Pope's Umbriel.

These three important figures—Belinda, Ariel, and Umbriel—inhabit Pope's poem; the other creatures in the painting are Fuseli's. On the far left a small male figure, prone with arms outstretched, zooms forward. He is identical to the fairy Cobweb (in *A Midsummer Night's Dream*) painted by Fuseli in the mid-1780s.[28] Perhaps he can be fitted into *The Rape of the Lock*, as a gnome inferior to Umbriel and of similar mischievous nature. (Just above him is the small, hardly

24. Schiff No. 681. In his *Lectures on Painting* (1801) Fuseli referred to these sculptures when he said that Michelangelo 'embodied sentiment on the monuments of St. Lorenzo' (p. 62).
25. Schiff No. 842. Fuseli used a similar pose for a dreaming woman as late as 1814 (Schiff No. 1496).
26. Schiff No. 467.
27. Schiff No. 750; also 751.
28. Schiff No. 752.

visible figure of a woman wearing a broad tilted hat and swirling in dance: she can be regarded simply as a Fuselian graphic doodle.)

Beneath the zooming figure sits a small scowling woman, balanced by a small smiling woman visible between the legs of Umbriel: these are the two aspects—malevolent and benign—of the fairy Queen Mab. As described in *Romeo and Juliet* (I.iv) she is the 'fairies' midwife', who 'gallops night by night/Through lovers' brains, and then they dream of love'. She thus fits the context of Belinda's dream, and connects it with Fuseli's most famous painting 'The Nightmare'.[29] The presence of Queen Mab again shows how Fuseli, when not confined to a literal illustration (as in 'The Cave of Spleen'), infuses Pope's poem with the imagery and poetry of Shakespeare—as though the freight of symbols in Pope's couplets was too slight, and needed to be made weightier with aid from the greatest English poet. To some extent, then, Fuseli has translated the Rosicrucian symbols of Pope's poem into the fairy lore of Shakespeare.[30]

In his way Fuseli was a poet in paint, for he added other elements to the composition that come from neither Pope nor Shakespeare. At the foot of Belinda's bed kneels a creature swathed in a sheet. Although this has been variously identified as the dog Shock about to awaken Belinda or as the Baron about to leap up to seize the precious lock[31] it seems far more likely to be a version of the witch-like hag in Fuseli's drawing called 'The Selling of Cupids'[32] (a paraphrase of a Pompeian wall painting). Although the precise meanings of Fuseli's symbols are not always clear, is there not a connection between Belinda's erotic dream and the seller of cupids waiting for her to awake?

That Belinda's dream is erotic seems clear enough when we examine what Fuseli has painted behind the crouching figure at the foot of her bed: a pair of white moths in the act of copulation, a

29. This connection is made by Kalman (n. 21 above). Fuseli painted several different versions of Queen Mab, though from Milton's 'L'Allegro' rather than Shakespeare (Schiff Nos. 909, 910, 1498).

30. The poet Francis Thompson also noted this connection between Pope and Shakespeare: 'the sylph machinery in "The Rape of the Lock" is undoubted work of fancy: the fairyland of powder and patches, "A Midsummer Night's Dream" seen through chocolate-fumes' ('Pope', *The Academy*, 3 July 1897, p. 13).

31. Schiff, I/I, p. 632 and in the Tate Gallery Exhibition catalogue of 1975, p. 107; Kalman, p. 227.

32. Schiff No. 655. It also resembles the crouching figure in the lower left of 'Percival Delivering Balisane. . . .' (Schiff No. 718). A similar hooded, witchlike figure, again next to one very like Umbriel, appears in a drawing of the same period (Schiff No. 1757).

repetition of the motif in his drawing 'Sleeping Woman and Cupid' (mentioned above, page 51). Derived from Swiss folklore, the moth was for Fuseli a constant dream symbol, present in his work as early as 1763.[33] A pair of moths locked in sexual union could only symbolize erotic dreams.

The painting as a whole goes far beyond Pope's poem, as we have seen, for it also embodies the products of Fuseli's far-ranging imagination when confronted by the poem—which served him as a spring-board into macabre and erotic regions of his own. William Hazlitt once described him as 'undoubtedly a man of genius, and capable of the most wild and grotesque combinations of fancy'.[34] 'The Dream of Belinda' bears this out. Why Fuseli painted it or what he hoped to do with it can only be a matter of speculation. He may have been discouraged from exhibiting it by what he regarded as the unsympathetic taste of the time for that kind of painting. He later told Joseph Farington, the diarist and painter, that he had 'little hope of *Poetical* painting finding incouragement in England. The People are not prepared for it. Portrait with them is everything.—Their taste & feelings all go to *realities*.—The ideal does not operate on their minds.—*Historical* painting, viz.: matter of fact, they may encourage.'[35] Fortunately his own energy and vision drove him to seize on Pope's poem for this '*Poetical* painting'. It invokes what Edgar Allan Poe described (with disapproval) as 'the certainly glowing yet too concrete reveries of Fuseli'.[36]

33. Schiff No. 353 and I/I, p. 435. The moths are probably the species popularly known as sphinx.
34. William Hazlitt, *Collected Works*, ed. A. R. Waller and A. Glover (1902–6), VII, 94.
35. In 1805, *Diary*, ed. J. Greig (1922–8), III, 91.
36. 'The Fall of the House of Usher,' *Complete Works* (New York, 1902; repr. 1965), III, 283.

V
Fuseli's Disciples

During his long career in England, Fuseli played a prominent role in academic art in spite of his eccentricity as artist and as personality. Elected a member of the Royal Academy in 1790 he rose to be Professor of Painting in 1799; and the publication of his lectures (in 1801–2) buttressed his reputation as teacher as well as practitioner. Under his inescapable influence three of his pupils illustrated *The Rape of the Lock*; and their drawings, although executed in the third and fourth decades of the nineteenth century, are permeated with the spirit and technique of their master (who died in 1825).

His favourite pupil was Theodore Matthias Von Holst, an Englishman in spite of his name (his father had emigrated from Riga).[1] He was astonishingly precocious, winning admission as a student at the Royal Academy shortly after 1820, when he was ten years old, and exhibiting a drawing there when he was seventeen. He became Fuseli's pupil soon after entering, and so strong was his master's influence on him that his works have often been attributed to Fuseli.

Like his master, Von Holst preferred romantic and gloomy subjects; he had a 'natural disposition inclining to melancholy' (his obituary states).[2] This would explain why, when he illustrated *The Rape of the Lock*, he preferred the Cave of Spleen as his subject. [*Plate 24*] While Fuseli's illustration of the same scene (for the 1798 edition of the poem) had shown Umbriel departing, Von Holst shows him (in the upper right corner) arriving at the Cave. But in other, more important ways Von Holst departed from his master's version. He enclosed the scene with a proscenium arch, as though it were on a stage, probably using as his model the Theatre Royal in Drury Lane; the self-caricature on the cornice (at the upper left)

1. The fullest account of this artist is Gert Schiff's 'Theodore Matthias Von Holst', *Burlington Magazine*, 105 (1963), 23–32.
2. *The Art-Union*, No. 64 (April 1, 1844), p. 87; quoted in Schiff, 'Von Holst', p. 24.

24. Theodore Matthias Von Holst, The Cave of Spleen [1820–30], watercolour

parodies the flying figure in the proscenium arch of that theatre.[3] The illusion of a stage setting is reinforced by the troubadour playing a guitar, his legs dangling into the orchestra pit—this is a self-portrait as signature—and by the scene in the background (on the right), a pavilion and grove of trees, which look like the painted backdrop of a stage set.

For some reason—not all of Von Holst's fantastic touches can be explained—he dressed Spleen as an oriental queen wearing headdress and ornaments. Her figure in general resembles the Cleopatra of William Etty, exhibited first at the Royal Academy in 1821, though (more probably) both queens had a common source in the Michelangelo sculpture that Fuseli had modelled his Belinda on (p. 50 above). Spleen is attended by four handmaids, of whom the two important, authentic ones—Affectation and Ill-Nature—are more prominently placed and more expressively painted. Von Holst carefully selected a choice group from the unnumbered throng of

3. *The Theatre Royal Drury Lane and The Royal Opera House Covent Garden*, Vol. XXXV of *Survey of London*, ed. F. H. W. Sheppard (1970), plates 27a, 32a, 33a.

grotesques listed in Pope's couplets: in the left foreground, the human tea-pot, the talking gooseberry pie, and (logically) an empty bottle to represent the maids calling aloud for corks. His most startling image, however, is the literal illustration of Pope's line: 'Men prove with Child, as pow'rful Fancy works', for the tiny figure (in the far right foreground) is a man, appropriately wearing a night-cap, who is clearly in the final stage of pregnancy.

Like Fuseli, Von Holst introduced into his composition figures not in the text: the jack-in-the-box head on the extreme right (above the pregnant man) may be a self-portrait; the dark, glowering demonic being (in the upper left) has a counterpart in Von Holst's obsessive illustrations to *Faust*; and the group of three white-hooded witches above the Queen may be a tribute to Fuseli, in whose works the three witches from *Macbeth* is a recurring motif. This fresh and original picture, a finished gouache done with evident care in the 1820s, was neither exhibited nor published as an engraving.[4]

William Lock the younger, another pupil of Fuseli's, was an amateur who did not need to struggle for his living, as Von Holst did; he was heir of a wealthy family whose seat was Norbury Park in Surrey, and he married a well-known beauty. Joseph Farington, who regarded his paintings as miserable in conception, greatly admired his drawings as 'exquisite', the equal of the best in the eighteenth century. Thomas Lawrence also thought highly of his talent, calling him 'Raphael William'. And his teacher Fuseli, who dedicated to him his lectures on painting, once pronounced a drawing of his 'Divine'.[5]

In the British Museum is an unsigned drawing of Belinda at her dressing-table, attributed to Benjamin West. [*Plate 25*] On stylistic grounds, however, it is more likely by William Lock, for its voluptuous grace is more characteristic of his work than of the Quaker West's.[6] It shows Belinda in a pensive mood and sitting in a sensuously curled posture. Above her head Ariel, mounted on a large butterfly, directs his legions of sylphs as they float and swirl about. Although Pope's verse assigns various duties to them, the artist has made a piquant addition: one happy sylph supports and embraces Belinda's exposed breast with an obvious pleasure far beyond his duty. The bat-like creature (in the upper left) is perhaps a gnome on a reconnaissance flight. On the floor a lapdog stretches out

4. Date of composition from Schiff, 'Von Holst', p. 27.
5. Duchess of Sermoneta, *The Locks of Norbury* (1940), pp. 26, 46, 220.
6. The few nude females that West painted are classic and chaste. See Grose Evans, *Benjamin West and the Taste of his Times* (1959).

25. [William Lock the Younger], Belinda at her Dressing-Table [n.d.], drawing

contentedly,—not the historically correct breed of shaggy, unkempt Shock but rather a svelte spaniel who is more in keeping with the sumptuous lines of the drawing.

The style of swirling elegance so characteristic of this drawing is even more fully exploited by Lady Georgina North. Her family were friends and patrons of Fuseli—particularly her grandfather Thomas Coutts, the banker, and her mother, the Countess of Guilford. Fuseli felt the strongest friendship for Lady Georgina, his friend John Knowles wrote; he had 'that affection which a master usually feels towards an amiable, accomplished, and highly promising pupil'.[7] In 1831 Lady Georgina drew a set of five pencil and watercolour illustrations for *The Rape of the Lock*, one for each canto. They would not have failed to please her teacher had he lived to see them, for they stand out as brilliant inventions executed with fine craftsmanship. The drawings were etched by W. Raddon but almost certainly they were not published to be sold, only to be presented to her family and friends.[8]

Her illustration for Canto I, Belinda asleep [*Plate 26*], immediately displays the pervasive influence of Fuseli in its highly wrought mannerism. The lilliputian figures, both sylphs and non-sylphs (who may be regarded as fairies), make the human-sized heroine look gigantic, a device exploited by Fuseli in his Shakespearean illustrations. The sylph hovering between the panels of the window curtain wears a large, tilted hat (again a Fuselian echo). With her ribbon she controls two doves, one placing a *billet-doux* on Belinda's dressing-table, the other perching on the edge of Shock's drinking basin.[9] Above the sleeper's head, directing her dream, is Ariel, crowned by a star to indicate rank. Although in the poem Ariel and his legions of sylphs are male, from Stothard's illustrations onward they were transformed into females. Perhaps associated in artists' minds with fairies, they became female, while the gnomes remained male. Since in the poem Ariel explains that the sylphs 'with ease/Assume what Sexes and what Shapes they please' (I. 69–70), henceforth for illustrators the sylphs were pleased to assume the female sex.

7. *The Life and Writings of Henry Fuseli* (1831), I, 341.
8. Her family's archives contain etchings of the drawings for Cantos 1, 2, 3, and 5 (Bodleian Library, North b.3, ff. 4–7).
9. In Lady Georgina North's numerous sketchbooks, now in the Bodleian Library, I could find only one drawing related to her series on *The Rape of the Lock*—a sketch of two doves on the edge of a basin with several sylph-like fairies nearby (North MSS, e.45, f.78).

26. Lady Georgina North, Belinda Asleep (1831), drawing

No previous illustrator had drawn the scene of the Baron's prayer, which Lady Georgina chose from the second canto of the poem. [*Plate 27*] The Baron, his long muscular legs outstretched in the Fuseli manner, prays for success in his rapacious mission. Dressed in doublet and hose, his plumed hat at his feet, he could be the leading dancer in a romantic ballet. Strains of *Giselle* or *Swan Lake* coming from the wings or orchestra pit would not be inappropriate. A large winged cupid, his tutelary deity, looks down on him solicitously. The lilliputian creatures at the bottom of the drawing are somewhat mysterious: on the right a woman sitting on an hour-glass plays a game of badminton using a winged creature as shuttlecock; beneath the Baron a winged man stands over a seated woman. The mischievous face, just visible behind the stand that supports the Baron's altar, is apparently a gnome, spying for Umbriel.

For the third canto [*Plate 28*], the climactic rape scene, Lady Georgina chose the moment when the Baron, who stands behind Belinda, has just snipped off the lock although she does not yet know it. The Baron has undergone a drastic change in features and clothes from the previous drawing: instead of the robust, muscle-bulging hero he is more delicately modeled, particularly his face, which at first glance may be mistaken for a woman's. The pair of scissors he holds in his right hand is still attached, by means of a long ribbon, to the person who handed it to him—the false Clarissa, who sits at Belinda's feet and looks up at her in treacherous friendship. Above the Baron's head hovers his attendant Cupid, while off to the side is Belinda's Ariel, identified by star and wand, powerless to help her mistress. On the extreme lower left we see the mischievous gnome again, this time about to prod the foot of the page-boy holding the tray in order to make him drop it. And on the lower right a curious little vignette parodies the main action: a man lifting from a woman's head a thatch of hair to display her bald skull. The butterflies on the left (below Ariel) are balanced by the diaphanous female figures on the right, while a sylph clutches the long ribbon to impede the Baron's rash deed. Behind the sylph another looks down in concern; she wears a large saucer-shaped Fuselian bonnet.

Although the most provocative scene in Canto IV is the Cave of Spleen, always chosen by those who had previously illustrated that canto, Lady Georgina preferred one that could be treated in a sentimental rather than satiric manner. When Umbriel returns from the Cave,

> Sunk in *Thalestris*' Arms the Nymph he found,
> Her Eyes dejected and her Hair unbound.

27. Lady Georgina North, The Baron's Prayer (1831), drawing

28. Lady Georgina North, The Rape of the Lock (1831), drawing

In the illustration [*Plate 29*] Umbriel is mounted on a branch of spleenwort, from which hangs the wondrous bag of sighs, sobs, passions, and war of tongues; and in his hand he carries the vial of fears, sorrows, griefs, and tears. Their contents will, when released, fuel Belinda's anger. Thalestris, as she consoles her, seems tender and compassionate, not yet hardened into the fierce, raging virago. At Belinda's feet lilliputian women mourn, while Shock looks up sympathetically at his desolated mistress.

For the last canto Lady Georgina again chose not the most dramatic scene—the battle to regain the lock—but rather the very conclusion of the poem. [*Plate 30*] But in the upper left corner she illustrated an earlier passage of the canto: 'Now *Jove* suspends his golden Scales in Air'—when Belinda's lock of hair outweighs men's wits. Here we see Jove flanked by his imperial eagles and tended by Hebe as he watches the scales. The main design is decorative rather than illustrative—a diaphanous swirl of sylphs and fairies carefully positioned on a large *S* curve as they pursue the lock transmuted into a comet in the heavens. Hogarth's Line of Beauty may be the remote source of this spiral trajectory, and William Blake's 'Jacob's Ladder' is also a composition of similar design; but the vibrant delicacy and cushioned elegance of Lady Georgina's drawing can be credited to the artist herself. With her fluent and fanciful style—it can be called Regency rococo—she creates a fantastic, fairy-tale atmosphere that blends Alexander Pope with Arthur Rackham.

Following the example of improvisation set by Fuseli, their master, these three artists—Von Holst, Lock, and Lady Georgina North—produced illustrations that are less faithful to the verbal text of Pope's poem than, say, Wale or Burney or Stothard. Yet, at the same time, endowed with fecund imaginations and facile pencils they caught the spirit of the mock-epic far more sympathetically than the literal illustrators.

One of the most curious, provocative illustrations evoked by Pope's poem came from the pen and brush of Charles Kirkpatrick Sharpe, the Scottish antiquary and amateur artist. [*Plate 31*] Characteristically satiric in his approach, he designed a clever amalgam of Pope's 'Cave of Spleen' (writing six lines of the text under the drawing) and Fuseli's most famous painting, 'The Nightmare'. The Fuseli, which was exhibited at the Royal Academy in 1782, had aroused sensational interest, and then through published engravings became known far beyond London art circles. It also became one of the most travestied paintings of its time and

29. Lady Georgina North, Belinda Grieving (1831), drawing

30. Lady Georgina North, The Lock Transmuted (1831), drawing

31. Charles Kirkpatrick Sharpe, The Cave of Spleen [n.d.], watercolour

throughout the nineteenth century. (Both Rowlandson and Cruikshank exploited it.) Generally it was adapted for political or personal caricature;[10] but Sharpe, unlike other artists who used it, linked 'The Nightmare' to a literary text, and emphasized the sexuality and scatology in both painting and poem. For that reason, probably, it remained unpublished even in the posthumous 1869 collection of his etchings.[11]

The most obvious travesty in Sharpe's grotesque composition is his combining the goddess Spleen with the dreaming woman of 'The Nightmare', while transforming Pope's Umbriel into Fuseli's incubus (or nightmare) sitting on the woman's breast. But Sharpe's parody goes beyond this particular painting of Fuseli's to echo other motifs from his *œuvre*. The faintly outlined sylphs floating above the candle—one of them a skeleton, a reminder of mortality—are intruders in Pope's Cave of Spleen, visitors from Fuseli's 'The Shepherd's Dream'; and the miniature bearded man in the lower left, from one of his Shakespeare illustrations.[12] In the upper left of the drawing the two figures brooding above the procession of old men

10. Nicolas Powell, *Fuseli: The Nightmare* (1973), pp. 17–18, 78–94, 101–3.
11. Charles Kirkpatrick Sharpe, *Etchings, with Photographs from Original Drawings, Poetical and Prose Fragments* (Edinburgh).
12. Schiff Nos. 1762, 753.

resemble the Fuseli witches from *Macbeth* summoning up the visionary procession of kings. (Two instead of three witches, and six instead of eight kings weaken this interpretation.) Sharpe added his own share of the fantastic. Facing Umbriel, another incubus squats on a chamber-pot, while on his head he wears a pot decorated with a plume. Why chamber pots? Perhaps a graphic symbol of Pope's phrase for a strange phantom in the Cave of Spleen: 'Lakes of liquid Gold'. Sharpe has furnished the table with a phallic support for the shrouded mirror (equivalent to the satyr's leg in the 1714 frontispiece), and has humanized the scissors by putting a head on it. In the lower left he has invented his own throng of fantastic creatures who inhabit the Cave. In its colouring, except for a few touches of bright red (particularly in the comb of the cock), the whole drawing is tinted with a wash of blue and green, very suitable to its splenetic subject.

VI
Romantic and Victorian Illustrations

The ranking of Pope as a superior craftsman of a secondary kind of poetry, initiated by Joseph Warton, became more marked as the tide of English romanticism rose in the early nineteenth century; and if *The Rape of the Lock* continued to win praise, its critics limited that praise with further qualifications. The exaltation of Nature (upper case) by Wordsworth and his school opened a gap so wide as to be unbridgeable between that and Pope's concern with human nature (lower case). William Bowles, in his notorious attack on Pope's character and poetry (in 1806), while extolling the poem as the best of its kind, then cautioned his readers to remember that it is 'founded on *local manners*, and the employment of the Sylphs is in *artificial life*; for this reason, the Poem must have a secondary rank, when considered strictly and truly with regard to its poetry'.[1] Yet in spite of himself Bowles was almost carried away by his enthusiasm for this poem 'in an inferior province of Poetry' when he confessed: 'we almost doubt whether the garb of elegant refinement is not as captivating, as the most beautiful appearance of Nature'.[2] To match elegant refinement against Nature was a great concession on Bowles's part, but the cautionary 'almost' saved him from abandoning his patronizing perch.[3]

Byron, who was Bowles's main antagonist, vigorously defended all of Pope's poetry as worthy of high esteem. And Isaac D'Israeli, another champion of Pope, replied more specifically to the denigration of *The Rape of the Lock* by asserting that the best poetry should represent the spirit of its own age. Judged by that standard, he continues, Pope can be classed with Dante or Milton, for was he not the preeminent creative genius of his own time? And hurling the

1. *The Works of Alexander Pope in Verse and Prose* (1806), I, 354.
2. Ibid., X, 373, 374.
3. Although Upali Amarasinghe calls the dispute about 'artificial' and 'natural' on the whole 'unenlightened and confused' (*Dryden and Pope in the Early Nineteenth Century*, Cambridge, 1962, p. 133) and René Wellek calls it 'puerile' (*A History of Modern Criticism*, New Haven, II, 1955, p. 123), the fact is simply true—that the juxtaposition of these two terms was a staple of Pope criticism until the end of the nineteenth century.

gauntlet at Pope's detractors, D'Israeli writes: 'it is equally probable that Dante and Milton, with their cast of mind, could not have so exquisitely touched the refined gaiety of "The Rape of the Lock".'[4] Far more typical was such a judgment as Thomas Campbell's (in 1808) that while the poem was the finest 'gem . . . in all the lighter treasures of English fancy' it could not claim the 'superior dignity' of the heroic creations of Spenser and Milton.[5]

Having firmly placed *The Rape of the Lock* on a secondary level of literary creation, a critic like William Hazlitt pointed to the particular elements of the poem that could not be matched in universal masterpieces. Describing it (in 1818–19) as 'the most exquisite specimen of *filigree* ever invented', he removed it from comparison with the great classics. 'It is made of gauze and silver spangles', he writes. 'The most glittering appearance is given to everything . . . Airs, languid airs, breathe around; the atmosphere is perfumed with affectation . . . It is the triumph of insignificance, the apotheosis of foppery and folly. It is the perfection of the mock-heroic!' His flattering discrimination effectively reduced the poem from a genuine literary masterpiece to an intrinsically inferior exercise. This judgment was reinforced by Leigh Hunt when he compiled an anthology of poetry entitled *Imagination and Fancy* (1844). 'Pope's paragon of mock-heroics would have been found in this volume,' he writes, 'but for that intentional, artificial imitation. . . which removes [it] at too great a distance from the highest sources of inspiration.'[6]

If such a critical attitude toward poetry prevented *The Rape of the Lock* from being illustrated in the exalted vein of a *Macbeth* or a *Paradise Lost* it could be accommodated in another style of painting that had become increasingly important since the late eighteenth century—genre painting. The travail of Belinda, albeit in an aristocratic setting, could be expressed by this school that relished emotional incident in a humble setting, all overlaid with sentimental varnish. That such an approach violated the satiric spirit of Pope's

4. *Quarterly Review*, XXIII (1820), 410.
5. *Edinburgh Review*, XII (1808), 77. The conclusion in Amarasingh's study of Dryden's and Pope's critical reputation from 1800 to 1830 is that it remained high, that in the periodicals Pope's defenders outnumbered his detractors. But this does not take into account the qualitative level of Pope's defence, that he was usually defended as a poet of a lower order.
6. William Hazlitt, *Collected Works*, ed. A. R. Waller and A. Glover (1902–6), V, 71–2; Leigh Hunt, *Imagination and Fancy; or, Selections from the English Poets*, A New Edition (1883), pp. v–vi.

poem disturbed its illustrators, apparently, as little as its readers: both accepted the poem in the spirit of their own rather than Pope's age. In various ways artists transformed Pope's mock-epic, domesticating it into something for their own times.

No doubt reflecting popular taste, nineteenth century artists who illustrated the poem chose Belinda as their subject rather than any other character or any scene in the poem. This can be attributed in part to 'woman worship', as it came to be called in the 1860s, when women were regarded as a source of moral inspiration to such an extent that John Ruskin declared that their virtue and wisdom redeemed men from weakness and vice.[7] As heroine Belinda could also arouse the kind of sympathy that sentimental Victorians found congenial. And so in book illustrations as well as in prints and paintings her contemporary image reached a large public of readers and picture viewers.

Richard Westall, who began his career as a history painter, turned to book illustrations about 1795 to become second only to Stothard in prolific output. (He could command higher fees, however, having been paid forty-five guineas for a drawing from the *Iliad*.[8]) For the title-page of the 1829 edition of *The Rape of the Lock, and other Poems* he provided a vignette [*Plate 32*] to illustrate the couplet printed beneath:

> Twas then Belinda if report says true
> Thy eyes first open'd on a billet doux.

While she reads the *billet-doux* and Shock jealously tries to capture her attention, her guardian sylph is still present—not the male commander of Pope's airy legions but clearly a female sustained by large butterfly wings and armed with a long wand. The heroine herself is so dowdy that it is hard to believe she could arouse the Baron to his rash deed, or to anything else. And for a belle accustomed to receiving flattering *billets-doux* she looks surprisingly startled. Her bed-alcove lacks aristocratic elegance; instead it has a distinctly middle-class atmosphere, more suited to *The Vicar of Wakefield*, the subject in fact of one of Westall's best known set of illustrations. At least Westall followed the text of Pope's poem. In the portrait 'Belinda' [*Plate 33*] Eliza Sharpe depicted a stolid young woman holding the unopened letter after she has been carefully dressed and her lock of hair temptingly arranged.[9]

7. Walter E. Houghton, *The Victorian Frame of Mind 1830–1870* (New Haven, 1957), p. 350.
8. Du Roveray to George Baker, 22 Jan. 1806 (BL Eg. MS 2679, ff. 13–14).
9. In 1827 Miss Sharpe exhibited 'Belinda' at the Royal Academy; in 1833 an engraving of it

32. Richard Westall and H. Rolls, title-page of Pope's *Rape of the Lock and Other Poems* (1828), engraving

33. Eliza Sharpe and H. Robinson, 'Belinda' (1836), engraving.

A lively Belinda was also designed by a different artist not to accompany the poem but as a picture to be exhibited and then copied as a print. Henry Fradelle, a French painter of historical subjects, had emigrated to London in 1816, and the following year began to exhibit in both the Royal Academy and the British Institution. Many of his pictures are based on literary subjects. In 1823 he exhibited at the British Institution 'Belinda at her Toilette'; and the following year, when it was engraved to be sold as a print, the proofs were first passed to him for his approval.[10] The print was evidently published in two sizes, the smaller one bearing the inscription, 'Engraved by permission of Mr. Fraddle [*sic*] from his large print'. [*Plate 34*] His portrait of Belinda almost fills the picture; and although he carefully furnished her dressing-table and provided a frisky lap-dog, a *billet-doux*, and an attentive maid-servant, he neglected to put into the picture the imaginative element of that episode in the poem: the sylphs who, rather than the maid, arrange Belinda's *toilette*. Both women are dressed in early eighteenth-century costume, but the décor in the room is of the artist's own day. Belinda herself is a well upholstered, matronly figure instead of the volatile, irresistible coquette invoked by the poem.

How far illustrators could depart in various directions from the spirit of Pope's mock-epic can be seen in the print entitled 'Belinda' by the artist-engraver Henry Moses. [*Plate 35*] He was patronized by Thomas Hope, the most energetic champion of the neo-classical revival during the Regency in England; and in his published engravings *Designs of Modern Costumes* (1812, enlarged about 1823) he used the furniture and decorations of Hope's famous London mansion as décor for his interior scenes.[11] In his print of Belinda he converts a Queen Anne coquette into a Regency courtesan who, like the sleeping Shock at her feet, is oblivious of the legions of sylphs, one of whom is so vigilant that he has even invaded the fish-bowl.

Most illustrators of the time sank in the swamps of sentimentalism. One such artist, Thomas Uwins, was a prolific painter who began to exhibit at the Royal Academy in 1803, when he was twenty-one, and until 1857 continued to show sentimental genre pictures,

appeared in *Heath's Book of Beauty* (with some anonymous sentimental verse unrelated to *The Rape of the Lock*), and in 1836 in *Heath's Book of Engravings* (with an extract from Pope's poem).

10. Letter from Fradelle, 30 June 1824, to 'Mr. White' [printer] in Brownlow Street (BL Add. MS 42 575, f. 283).

11. David Watkin, *Thomas Hope and the Neo-Classical Idea* (1968), p. 52.

34. Henry Fradelle and J. Phelps, Belinda at her Dressing-Table (1824), engraving.

many of them set in Italy, where he had lived for seven years. Early in his career, about 1808, he began to design frontispieces and vignettes for books. His portrait of Belinda, in an 1835 edition of Pope, shows her framed by an arch of sylphs, while in the lower background a pair of eighteenth-century beaus and a pair of urns serve as historical setting. [*Plate 36*] His other illustration for the poem is the vignette on the title-page of Pope's *Poetical Works* published in New York in 1828. [*Plate 37*] Its title, 'The Whisper in the Dark', is a phrase from the first canto of the poem, but wrenched from its context there by the mood of the drawing. We see a pair of lovers—probably intended to suggest Belinda and the Baron; both are dressed and posed as though playing Romeo and Juliet melting into each other's arms. From above, a sylph has come close to the heroine's ear, advising or warning her, while the clump of sylphs, with arms raised, seem to be lamenting the romantic union.

A variation of the sentimental approach can be seen in an

35. Henry Moses, 'Belinda' [1823?], engraving

engraving of Belinda reading her *billet-doux* [*Plate 38*] by Wilson Dyer, an obscure painter from Manchester, who exhibited several times in the British Institution in the 1850s. But he adds still another distortion: Belinda's maid, who was only praised for 'Labours not her own' and plays a negligible part in the poem, has been converted into an active character as she peers over Belinda's shoulder to read the lover's message. The expression on her face, in contrast to her mistress's thoughtful and sentimental satisfaction, is one of anxious, perhaps malevolent, glee.

A more congenial illustration for the poem was designed by William Witherington as the frontispiece for an edition of Pope's works in 1835. [*Plate 39*] Mainly a landscape painter, Witherington frequently exhibited at the Royal Academy for more than half a century. His human figures are generally well drawn. The scene of the poem that he chose to illustrate is the climactic one of the rape itself. His sylphs are tiny gauzy creatures, hardly visible. As the Baron prepares to cut the lock, one sylph is twitching Belinda's diamond earring to warn her of the approaching danger. Another sylph, below, throws up her hands in despair; it is Ariel, apparently, whose power expires as he views an earthly lover lurking in Belinda's

36. Thomas Uwins and H. E. Shenton, Belinda [n.d.], engraving

37. Thomas Uwins and Sceles, 'The Whisper in the Dark', Pope's *Poetical Works* (1828), engraving

38. Wilson Dyer and C. Rolls, Belinda's *Billet-doux* [1850–60], engraving

39. William F. Witherington and A. W. Warren, frontispiece to Pope's *Works* (1835), engraving

heart. The humans in this tableau, especially in contrast to the gauzy sylphs, are phlegmatic and stolid, almost monumental in their weightiness and their static pose.

In contrast to this interpretation of the rape-scene is the frontispiece [*Plate 40*] to an 1860 edition of Pope's *Poetical Works* by one T[homa]s Brown.[12] The scene is frenetically agitated as the Baron holds up the lock of hair, while Belinda and Sir Plume look on in horror. The cluttered excitement is matched by its Victorian decoration; antimacassar embroidery covers everything, from Belinda's footstool and lapdog to the top of the mirror in the background. The style of the composition seems to us today a mixture of low kitsch and high camp.

At about the same time that publishers were sponsoring this kind of hackwork as illustrations for their books, an independent artist of stature found genuine inspiration in Pope's classic. Charles Robert Leslie painted one of the most distinguished interpretations of the poem between those of Fuseli and of Beardsley—distinguished for historical accuracy and artistic excellence. He designed it not as a book illustration but as an easel picture commissioned by a wealthy collector.

Early in his career C. R. Leslie was convinced of the importance of acquiring knowledge of literature. 'I am now going through Homer, Milton, and Dante's works, which every painter should be well acquainted with', he wrote to his sister in 1813. During his long, busy career he found his subject matter in Shakespeare, Cervantes, Molière, but mainly in eighteenth century writers: Le Sage, Addison, Fielding, Goldsmith, Smollett, and Sterne.[13]

Whether the notion of using Pope's most popular poem arose in his own mind or was suggested by the collector who commissioned the picture, in May 1851 he reported to his sister that he was 'very busy with a large picture from "The Rape of the Lock" for Mr. [John] Gibbons, and am indeed overwhelmed with commissions'. By May 1852 he had not finished it in time for that year's Royal Academy exhibition. He spent the summer months of August and September at Hampton village, and from there informed his sister that he was painting the background 'perhaps in the very room where the scene of my picture occurred'. He was more expansive in

12. He is not easily identifiable—perhaps the painter and sculptor who exhibited at the British Institution: a painting in 1854 and a sculpture in 1855, both on Biblical subjects.

13. Charles Robert Leslie, *Autobiographical Recollections*, ed. Tom Taylor (1860), II, 38–9, 44.

40. Thomas Brown, frontispiece to Pope's *Poetical Works* [1860], engraving

41. Thomas S. Seccombe, Belinda Embarking, Pope's *Poetical Works* [1878], engraving

writing (at the end of August) to his friend the American writer Washington Irving: 'I am painting a large picture from "The Rape of the Lock", containing fourteen or fifteen figures. I have taken the moment in which Sir Plume is desiring the Baron to return the lock. Belinda is in the foreground crying, and surrounded by ladies, and the group of gentlemen further in the picture. As the back ground represents a room in the palace, I am finishing the picture there. I can paint at the palace for two or three hours each morning, uninterrupted by visitors, and on Friday, when it is closed to the public, the whole day.'[14]

In one of the small oil sketches that he painted he blocked in the main groups of figures and something of the setting.[15] Although illness prevented him from completing the final version of the picture in time for the Royal Academy show in 1853, it was privately seen and generously praised by a magazine critic as 'perhaps the most exquisite work which the genius of its author—fertile as it is—has ever yet given to the world'.[16] Finally, the large finished picture was exhibited in 1854 under the title 'Sir Plume demands the Restoration of the Lock'. [*Plate VI*][17] It has apparently never been engraved.

As described by Tom Taylor, Leslie's friend and editor: 'The scene represents the moment when Belinda mourns over the discovery of the ravished lock. She is weeping in the foreground, surrounded by a sympathetic group of ladies. The Amazonian Thalestris [centre], in *tricorne* and riding habit,[18] indignant at the peer's boldness, grasps her whip with an evident longing to use it over the insolent beau's shoulders. In the background Sir Plume [with cane and open snuff-box] is occupied on his unavailing mission, and the peer displays the captured lock in triumph . . . As a composition this is among the best works of Leslie's pencil, though there is an unpleasant predominance of that chalkiness in colour

14. Ibid., II, 301, 305, 306. For Leslie as an 'artist-antiquarian' (of historical scenes) see Roy Strong's *And when did you last see your father? The Victorian Painter and British History* (1978), pp. 62, 123–6.
15. Its size: 9 by 13 inches; now in the Tate Gallery.
16. *The New Monthly Magazine*, ed. W. H. Ainsworth, 98 (1853), 32–4.
17. It is uncertain whether this is the 1874 picture or the version Leslie painted for Edwin Bullock two years later.
18. It would seem a breach of decorum for a lady to attend a reception at Hampton Court in her riding dress. At Bath, under the benevolent despotism of Beau Nash, few ventured to appear at the assemblies in riding-dress; and if a gentleman, through ignorance or haste, appeared in boots, Nash would tell him that he had forgotten his horse (Oliver Goldsmith, *The Life of Richard Nash* in *Collected Works*, ed. A. Friedman, Oxford, 1966, III, 306).

which grew upon him during the last ten years of his practice . . . The Sir Plume is as genuine as the [Baron] is unreal. The tall and commanding lady in the crimson sacque, whose back is turned to the spectator in the foreground, is a masterly example of drawing and colour, and the picture is deserving of close study by young artists for the great art shown in its easy, natural, and yet most profoundly calculated composition. It is a capital example, too, of Leslie's admirable management of light and shadow'.[19]

As characters in the picture Leslie used models who in a few instances can be identified. To impersonate the Baron he prevailed upon his friend John Everett Millais, already well known as a painter and about to be elected to the Royal Academy. He first painted Millais's portrait on a small panel just as he was, wearing a black frock coat and a black cravat with a little golden goose for a pin. It was a very good likeness of him.[20] He then transferred the portrait to the large painting, dressing Millais in eighteenth century costume. In the opinion of Tom Taylor, the Baron was 'the weakest figure in the composition. Strange to say he does *not* look like a gentleman of the time of Pope, but like a modern gentleman masquerading.' And in the group around the aggrieved Belinda, Leslie painted in two of his own daughters.[21]

In his determination to make the setting as authentic as possible, Leslie had not only painted the picture in the very room in Hampton Palace where he presumed the incident had occurred almost one hundred and fifty years before, but he based his drawing of the furniture on the eighteenth-century collection at Petworth, the country seat of Lord Egremont in Sussex. Until Lord Egremont's death (in 1837) Leslie had visited Petworth almost every year for more than ten years, staying between one and two months each time.[22] (Egremont extended his patronage by buying eight of Leslie's paintings for his collection.[23])

In his interpretation of *The Rape of the Lock* Leslie enlarged the particular episode he illustrated. Thalestris foretells (in Canto IV) that with the Baron in possession of the lock of hair Belinda's reputation will be gossiped away; she will become a 'degraded

19. Leslie, op. cit., I, lxiv–lxv.
20. John Guille Millais, *The Life and Letters of Sir John Everett Millais* (1899), I, 164. This 1852 portrait of Millais is now in the National Portrait Gallery.
21. Leslie, op. cit., I, lxiv–lxv.
22. Ibid., I, lxiv, 162–3.
23. Richard Walker, 'The Third Earl of Egremont, Patron of the Arts', *Apollo Magazine*, Jan. 1953, p. 12. As a patron Egremont is better known for his hospitality to Turner.

VI. C. R. Leslie, 'Sir Plume Demands the Restoration of the Lock' (1854), oil

Toast'. No fewer than three groups in the painting exemplify the gossips: the couple on the far right, the two men in front of the fireplace, and the group of men in front of the window—including the ravisher himself, displaying the hair. Leslie has also subtly indicated how the Baron was able to cut Belinda's lock without warning. For the Baron must have been sitting in the upholstered armchair, now vacant, on which she leans; and when she had turned her back to him in order to bend her head 'o'er the fragrant Steams', he had snipped off the lock. Then, in hastily making his escape, he had dropped his hat, which still lies on the carpeted floor in front of Belinda. In her distress she has placed her coffee-cup, its spoon projecting, onto the seat of the chair. In the lower right we see Shock—authentic as to breed—lapping his afternoon tea, oblivious of his mistress's plight. It is indeed a remarkable historical reconstruction, for Leslie captures much of the authentic setting through costumes, furniture, woodpanelling, crystal chandelier, Chinese screens, Negro manservant. It all has the atmosphere of the time of Queen Anne, who in a full-length portrait (by Godfrey Kneller) in the background looks down goddess-like on the petty human scene being enacted below.

When the *Art-Journal* reviewed the Royal Academy exhibition that year its critic praised Leslie for the curious reason that he had not depicted 'sylphs and sprites': 'it is very certain that their presence upon canvas would be an impertinence, amid a society assembled at Hampton Court . . . Whatever Fuseli and Stothard may have done in their interpretations [in 1798], the author of the present work defers to the taste of the day.' In deferring to the taste of the day Leslie was reflecting the Victorian scientific assumption that—as a historian has put it—'at long last the human mind, freed from the false methods of theology and metaphysics, possesses the key to truth'.[24] But it was not the key to the poetic truth of Pope's imaginative use of Rosicrucian machinery; and in disregarding that element Leslie converted the comedy of mock-heroic epic into drawing-room tragedy.

Although the *Art-Journal* critic found fault with the execution of the painting, he concluded with high praise: 'We doubt indeed if [the artist] ever produced a picture better than this in all the loftier essentials of Art'.[25] A far more important critic, John Ruskin, paid

24. Houghton, *The Victorian Frame of Mind 1830–1870*, p. 33.

25. *The Art-Journal*, VI (1854), 162. Much of this review was lifted, without acknowledgment, by James Dafforne in *Pictures of Charles Robert Leslie* [1875], p. 10.

the picture the ultimate compliment as 'an absolute masterpiece, and perhaps the most covetable picture of its kind which I ever remember seeing by an English artist. Equal to Hogarth in several of its passages of expression, it was raised in some respects above him by the exquisite grace and loveliness of the half-seen face of its heroine, and by the playful yet perfect dignity of its hero. Nor was it less admirable as a reading of Pope; for every subordinate character had been studied with such watchful reverence to every word in which it is alluded to throughout the poem, that it seemed to me as if the spirit of the poet had risen beside the painter as he worked, and guided every touch of the pencil.'[26] How could Leslie have caught the spirit of Pope's poem so perfectly as Ruskin says when he has omitted such an essential element—the sylphs? One must rather regret that Leslie had deferred to the spirit of his own age.

Having pleased one patron with his picture, Leslie painted another version of it two years later for another collector, Edwin Bullock, who lived near Birmingham. One of the changes he made in the picture was to introduce portraits of Bullock's daughters in place of his own. This picture, however, he did not send to be exhibited. The patronage of such wealthy private collectors sustained mature artists who needed time to produce their work. Leslie explicitly acknowledged this while working on *The Rape of the Lock*, when he had ahead of him ten years of commissioned works (at his own price). 'The increase of private patronage of Art in this country is surprising', he wrote. 'Almost every day I hear of some man of fortune, whose name is unknown to me, who is forming a collection of the works of living painters. . . .'[27] But those artists whose reputations were not substantial enough to attract collector-patrons had to find employment as illustrators, for periodicals as well as books.

In the 1860s a group of brilliant illustrators (including the Pre-Raphaelites) exhibited their work in books and magazines;[28] but neither they nor their publishers were moved to illustrate *The Rape of the Lock*. Instead, when the poem was reprinted among Pope's works, second or third rate artists were hired to produce the vignettes or plates so common in books published during that period. Thomas Brown's 1860 illustration (p. 77 above) is one example. In 1878 Pope's poetical works, edited by William Michael Rossetti, brother of the poet-painter, contained an inferior illus-

26. 'Academy Notes, 1855', *Works*, ed. E. T. Cook and A. Wedderburn (1904), XIV, 38.
27. Leslie, *Autobiographical Recollections*, I, lxiii–lxiv; II, 302–3.
28. Gleeson White, *English Illustration 'The Sixties' 1855–70* (1897; repr. Bath, 1970).

tration for the poem. [*Plate 41*] It shows Belinda boarding the 'painted vessel' to sail to Hampton Court; she strides ahead—not a coquette, but rather like the lady of the manor on her way to open a village fête. In the background the men, in more distinctive eighteenth-century costume, look as though they are dressed for a masquerade. The artist was Colonel Thomas S. Seccombe, who exhibited in London from 1876 to 1885. He was a military painter, illustrator of children's books, and occasional contributor to magazines. His illustration also reminds us that the Victorian woman was not regarded as Venus-like; her 'sexual attraction was kept under wraps, many and voluminous'.[29]

The art of illustration, when performed as hackwork, could sink to a very low level. Another striking example is the crude woodcut by one George Kirby entitled 'Belinda's Toilet', published in a magazine in 1875. [*Plate 42*] It was printed on a page facing the poem 'Belinda's Toilette' by T. H. S. Escott, a journalist, who contrasts Pope's Belinda with modern belles, who veil their *toilette* from 'the prying male' and only when fully dressed 'burst upon a cynic world'.[30] This poetaster had forgotten that in Pope's poem Belinda does not hold a morning levee; she has only her maid and the invisible sylphs with her.

The explanation for such inferior illustrations as this magazine woodcut was suggested by an anonymous contributor to the *Quarterly Review* (in 1873) who, after complaining of the sad state of English painting, continued: 'The deterioration of Art among us is in some measure also due to the number of drawings continually in preparation to be poured from the press in the shape of cuts for our periodicals, newspapers, and illustrated books. These are generally required to be done on the spur of the moment, allowing no time for the completion of a well-digested design, so that the artist, to assist the imagination, or rather to find a substitute for it, is compelled to summon the aid of models or sitters before he has the least notion of what he wishes to say, and by their various arrangement and combination to adapt himself to every occasion. Of course, this is quite fatal to every valuable quality in Art.'[31]

The nineteenth century, then, was not a period when Pope's mock-epic elicited book illustrations of great distinction. The Victorian temper was too preoccupied with grand and profound

29. Houghton, p. 353.
30. *Belgravia*, Third Series, VII (1875), 325–6.
31. *Quarterly Review*, 134 (1873), 294.

42. George Kirby and C. M. Jenkin, 'Belinda's Toilet', *Belgravia* (1875), engraving

questions. 'To play with words was just as shameful as to play with ideas', a modern historian has written. 'The style of wit, paradox, and epigram, so characteristic of eighteenth century taste and so natural for a dandy, became intolerable.'[32] Many Victorians found the poem 'artificial'. That accusation, so argued about at the beginning of the century, continued to exercise critics, including the Rev. Malcolm Elwin, Pope's editor (in 1871), who quoted Bowles' strictures.[33] Perhaps in reply to this, W. M. Rossetti began the Prefatory Notice in his edition of Pope with a defence: 'A poet of an artificial age, and of artificial life, who is truly a poet, is a possession to be proud of: England can claim in Pope such a poet of her own.'[34] It was not a concept that would appeal to artists and critics 'hostile to any literature not seriously—even solemnly—concerned with fundamental questions'.[35]

32. Houghton, p. 225.
33. *Works*, ed. W. Elwin and W. J. Courthope (1871–89), II, 119.
34. *Poetical Works*, ed. W. M. Rossetti [1878], p. [xiii].
35. Houghton, p. 227.

VII
'Embroidered by Aubrey Beardsley'

Almost two hundred years after its initial publication *The Rape of the Lock* inspired an artist to invent a set of drawings that have become the most famous of all. In choosing to celebrate Pope's masterpiece Beardsley shared the critical taste of the *fin de siècle* rather than the disapproval that prevailed in the 1880s when, still in print, the poem had sunk to low repute. One popular anthologist (in 1880), after granting Pope the virtue of having treated contemporary manners, nullified his praise: 'because the incidents are trivial and the personages contemptible, Pope is not more than pretty in *The Rape of the Lock*.'[1] More influential critics also denigrated the poem. Matthew Arnold, that same year, cast a stern eye on Pope: his pronouncement became a classic of literary criticism. Applying his 'touchstone' method of judging poetry—matching poems against the acknowledged masterpieces of world literature—he observed a severe lack in Pope ('the high priest of an age of prose and reason') because his 'criticism of life' did not possess 'a high seriousness, or even, without that high seriousness, . . . poetic largeness, freedom, insight, benignity'.[2]

Leslie Stephen, far more involved in eighteenth-century literature and philosophy, had called the poem 'a masterpiece of delicate fancy', but later moderated his praise because of Hippolyte Taine's condemnation of the poem, and called it merely 'admirable after its kind'.[3] In his biography of Pope for the English Men of Letters series Stephen expanded his criticism. *The Rape of the Lock*, he wrote, burlesques a heroic style now so 'hopelessly effete' that it has lost its point: 'Pope's bit of filigree-work, as Hazlitt calls it, has become

1. *The English Poets*, ed. T. H. Ward (1880: rpt. 1891), III, 59. Yet in the 1870s and 1880s Pope was one of the ten most favoured poets used in school books, though low among the ten (Richard D. Altick, *The Common Reader A Social History of the Mass Reading Public 1800–1900*, Chicago, 1957, p. 161 n.).
2. *English Literature and Irish Politics*, ed. R. H. Super (Ann Arbor, Michigan, 1973), p. 180.
3. *Cornhill Magazine*, 28 (1873), 584; *Hours in a Library* (1907), I, 130–1.

tarnished'; and in spite of its 'dazzling display of true wit and fancy', it lacks that 'real tenderness and humour which would have softened some harsh passages, and given a more enduring charm to the poetry'.[4]

But to *avant-garde* writers and artists the poem seemed a world elegantly attenuated and refined, deliciously rococo, concerned with Art and the artificial rather than with Nature and the natural; they found utterly congenial what mid-Victorian taste rejected. The notable exception among *fin de siècle* writers, Oscar Wilde, wittily remarked that 'There are two ways of disliking poetry. One is not to like it, and the other is to read Pope'.[5] But that was not the opinion of the nineties, Vincent Sullivan has written. 'Dowson had an intense admiration for Pope, so had Lionel Johnson, so had Beardsley . . .'[6] By then Edmund Gosse had vigorously sprung to the defence of *The Rape of the Lock* against the strictures of Leslie Stephen, whom he accused of unfairness in 'breaking such an exquisite butterfly of art on the wheel of his analysis'. In Gosse's own opinion, 'Poetic wit was never brighter, verse never more brilliantly polished, the limited field of burlesque never more picturesquely filled, than by this little masterpiece in Dresden china.'[7] Gosse's judgement was consequential, for it was at his suggestion that Aubrey Beardsley came to illustrate *The Rape of the Lock*.

As Gosse recalled some seven years later, 'when A. B. was doing so much illustrating work of a trivial kind, in 1894–6, I repeatedly urged upon him the idea of illustrating, not ephemeral works of the day, but acknowledged masterpieces of old English literature, which should be in his own spirit. He replied, in his eager, graceful way, "Set me a task,—tell me what to do and I will obey." I said, "Very well, I will set you a task. You shall illustrate Pope's 'Rape of the Lock', Ben Jonson's 'Volpone', and Congreve's 'Way of the World'." He replied that he would certainly do the "Way of the World", which "jumped to his eye", but that he did not "see" the other two. No more was said, and I thought he had forgotten all about it, when I received "The Rape of the Lock", dedicated . . . to me.'[8]

It would seem to be an odd choice for an artist whose reputation

4. *Alexander Pope* (New York, 1880), pp. 40, 42.
5. In *De Profundis* he again denigrated Pope's poetry (*Letters*, ed. Rupert Hart-Davis, 1962, p. 482).
6. *Aspects of Wilde* (New York, 1936), pp. 161–2.
7. *A History of Eighteenth Century Literature (1660–1780)* (1889; New York, 1929), p. 113.
8. Gosse to A. E. Gallatin, 19 June 1902, MS in Princeton University Library.

was summed up by the *Westminster Review* in its notice of his collected drawings: 'everywhere the cynic leer of *fin-de-siècle* decadence'.[9] Beardsley's work for *The Yellow Book* as well as his association with Oscar Wilde's circle and his own dandyism had made him, in the public eye, the exponent and exemplar of wit, elegance, and depravity. Still a very young man—for he was born in 1872—he had by then absorbed the influence of the Pre-Raphaelites (in his *Morte d'Arthur* illustrations), of Japanese prints and Greek vase-painting (in his *Salomé* drawings); and he was beginning to display the influence of French eighteenth-century art, both decorative and illustrative.[10] It is not so strange, then, that he should find a congenial subject in the 'little masterpiece in Dresden china'.

Besides, his interest in eighteenth-century literature was of long duration. William Rothenstein, who knew him, remarked that he had 'explored the courts and alleys of French and English seventeenth and eighteenth century literature'.[11] At Brighton Grammar School he had read Jonathan Swift's *A Tale of a Tub* and *The Battle of the Books* with intense interest, and sketched on the book's flyleaves and in its margins various illustrations, including a full length sketch of Swift (looking more like a fop than a gloomy clergyman).[12] Perhaps Beardsley was receptive to Gosse's suggestion that he illustrate Congreve because as a schoolboy of sixteen he had drawn a scene from *The Double Dealer*, a conventional if gauche costume plate, with only its prominent display of candles to hint of his mature obsession.[13] Ultimately, though, he never illustrated *The Way of the World*; and he did not begin on *Volpone* until shortly before his death, leaving it unfinished. At the end of 1895, however, he 'saw' *The Rape of the Lock* with such clarity that he was able to complete its ten illustrations (including the cover) in a few months.

For one whose previous successes had been Malory's *Morte d'Arthur* and Wilde's *Salomé*, Beardsley would need a publisher who believed he could succeed with a literary work so different from these. Fortunately, in Leonard Smithers he had found a champion and a patron. Smithers, formerly a solicitor in Sheffield, had moved to London in 1891 to set up in his new trade. Although his

9. *Westminister Review*, 147 (1897), 596.
10. John Rothenstein, *The Artists of the 1890s* (1928), pp. 171–2.
11. *Men and Memories Recollections. . . . 1872–1900* (1931, 1939), I, 136.
12. Reproduced in W. B. Carnochan, 'Swiftiana: Beardsley's Illustrations of Swift', *The Scriblerian*, IX (1976), 57–60.
13. *Uncollected Work*, ed. C. L. Hind (1925), plate 101.

publishing career lasted only nine years, it is remarkable that with such precarious financial resources as he had he should have published some of the most important books of that decade. His motto was 'I'll publish anything that the others are afraid of.' When the scandal of the Oscar Wilde trial induced the publisher John Lane to dismiss Beardsley from *The Yellow Book* Smithers offered the young artist £25 a week for his exclusive services as co-editor (along with Arthur Symons) of a new magazine to be called *The Savoy*.[14] The prospectus for the magazine, dated November 1895, sets forth the credo of its editors: 'We have not invented a new point of view. We are not Realists, or Romanticists, or Decadents. For us, all art is good which is good art.' Significantly, the first issue of the magazine, in January 1896, contained the prospectus for *The Rape of the Lock* and an order form for its purchase.

Headed '"The Rape of the Lock" by Alexander Pope Illustrated by Aubrey Beardsley', the prospectus states: 'Mr. Leonard Smithers begs to announce the publication, in February next, of an Édition de Luxe of the above famous Poem, printed in crown quarto size, on good paper, illustrated with eight elaborate drawings by Mr. Aubrey Beardsley, and bound in a pictorial cover.' The edition, it continues, was limited and priced at 10/6, with twenty-five copies printed on Japanese vellum costing two guineas. In view of Beardsley's wretched health—he suffered from a virulent tuberculosis that would take his life two years later—he was brave to accept the job of producing the large set of illustrations for the poem while he was also working for the magazine. And it was bold of Smithers to promise subscribers the edition when the illustrator had barely begun his task.

On 5 January 1896 Beardsley attended the party at which Smithers launched *The Savoy*. At about the same time he wrote to a friend that he 'was just beginning some pictures for an edition of the *Rape of the Lock*'. In February, to escape from London's brutal weather he moved to Paris, which he found suited him very well, and there he industriously continued with his work. When he sent Smithers the sixth drawing—the 'Rape' scene—he wrote: 'The meticulous precision and almost indecent speed with which I have produced the drawing I send you by this post will prove to you that I have been nowhere near the Rue Monge [on the Left Bank]. I remain the same

14. George Sims, 'Leonard Smithers A Publisher of the Nineties', *London Magazine*, III (1956), 33, 35.

old hardworking solotaire you know and love so well. The drawing is one of the best I have done for the *Rape*.'[15]

Before long (on 7 March) he could tell Smithers: 'I sent you yesterday the "Cave of Spleen" drawing [,] which was completed amid the distractions of toothache & diarrhoea. Still!!'[16] Smithers showed his appreciation with a cheque and praise of the drawing; and by return post Beardsley promised to send the last full page drawing—the Battle of the Beaux and Belles—in a day or so. 'I think the sylphs and the cul de lampe had better be thrown into one,' he continued, 'and form the terminal decoration to the book. In all, that will be nine drawings, plus the cover.' True to his pledge he sent off the full page, and promised the *cul-de-lampe* the next day. (But neither here nor in any of the other drawings for the book are the sylphs depicted.)

Obviously Smithers was trying to hasten the production of the book, already delayed beyond its promised publication date. He had sent Beardsley the book's title-page, on which the word 'Illustrated' had been changed to 'Embroidered'; Beardsley thought the page 'quite charming'.[17] He also approved of Smithers' suggestion that one of the drawings be printed in the second issue of *The Savoy* in April; and although he chose the 'Battle' plate, which he called 'the last and best', Smithers preferred to use the sixth, the 'Rape',[18] probably because its title was the same as the book's. About 12 March Beardsley wrote to his sister, '*The Rape of the Lock* is finished and I think with great success.'

The second issue of *The Savoy* still needed to be put together, and that, he told his sister, would keep him 'dreadfully hard at work'.[19] Although his name was not listed on its cover as co-editor—Symons's name stood alone—he contributed Chapter IV of his fanciful novella 'Under the Hill', the first three chapters having been printed in the first issue; and the drawing of a rococo boudoir scene that decorated most of the cover was boldly signed with his name. Among the books advertised at the back of the magazine was *The Rape of the Lock* 'illustrated with nine elaborate drawings . . .bound

15. *Letters*, ed. H. Maas, J. L. Duncan, W. G. Good (1970), pp. 110, 115 ('solotaire' from Stanley Weintraub, *Aubrey Beardsley Imp of the Perverse*, University Park, Pennsylvania, 1976, p. 181).
16. *Letters*, p. 116 (corrected from MS in Huntington Library).
17. *Letters*, pp. 117, 118. But the spine of the book retained the word 'Illustrated'.
18. *Letters*, p. 124.
19. *Letters*, p. 118.

in a specially designed cloth cover'; and after directing the reader to the 'Specimen Drawing' on page 111 the notice concluded emphatically that the book was 'NOW READY'.

The binding that Beardsley designed is such a masterpiece of art nouveau that he justifiably put his initials in the bottom corners of its inner frame. [*Plate 43*] Its symmetrical design, stamped in gold on turquoise blue cloth, is a sumptuous and provocative introduction to the poem within. (The binding of the *de luxe* edition, gold on off-white, is bland and indistinct.) On a profane altar (or simply the lower border of a rectangular frame) stand a pair of elaborate candelabra, whose many candles are unlighted. Their stems and bases are ornamented with sinuous swirls of vegetable forms, though if examined more closely these take on a resemblance to shapes of animals or humans or foetuses. The candelabra support between them an oval frame for a pair of scissors above which floats the lock of hair, its curved form like a flame or a spermatozoon or a tiny animal (perhaps a sea-horse). The outermost border of the design is lined with a continuous succession of curves and bulges that could be lips or breasts or buttocks, or all of them. The binding, with its symbolism and ambiguities, thus serves the same function as the frontispiece in Du Guernier's 1714 illustrations by synoptically preparing the way for the drama within. As one critic has remarked, 'Beardsley's bindings are . . . maps of the perverted world inside the covers.'[20]

Once past the binding, so insistently *fin de siècle* in style, the reader is confronted by a title-page as insistently *dixhuitième* in its type-face and its use of black and red inking. [*Plate 44*] On the title-page of the 1714 version of the poem Pope had put a brief Latin quotation from Ovid, which he later replaced with one from Martial; on Beardsley's title-page Smithers generously and justly printed both quotations, for both are appropriate.

The frontispiece [*Plate 45*] is puzzling. Its title—like all the others, supplied by Beardsley—is 'The Dream'. An elegant courtier, carrying a long baton tipped with a glittering star, peers through an aperture in the bed-curtains. If Belinda is within, she is out of sight. Probably the courtier represents the disguised Ariel from Belinda's morning dream in the line: 'A Youth more glitt'ring than a *Birth-night Beau*'. A more persuasive explanation of the puzzle can be detected in the drawing entitled by Beardsley 'The Impatient

20. Peter Conrad, *Times Literary Supplement*, 25 March 1977, p. 336.

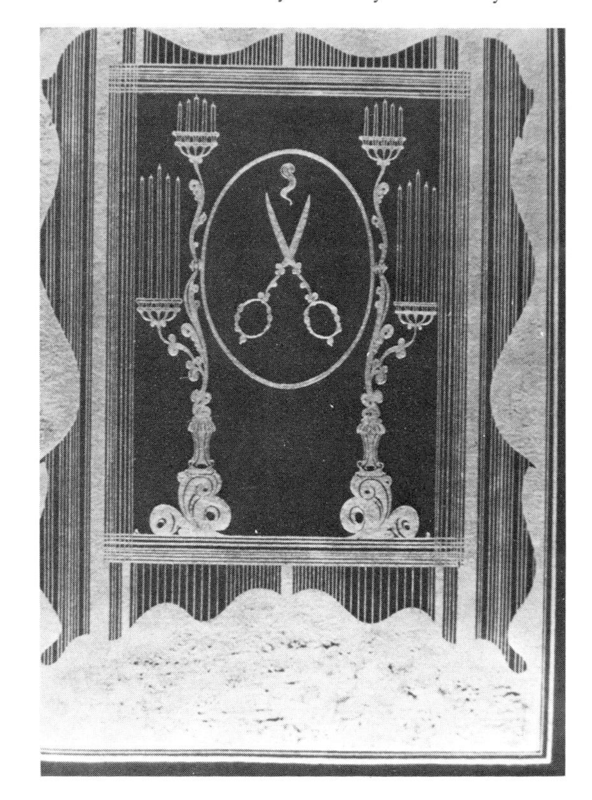

43. Aubrey Beardsley, front cover of *The Rape of the Lock* (1896), stamped cloth

Adulterer'. [*Plate 46*] In the summer of 1896 he was translating and illustrating Juvenal's sixth satire, in which a brief passage describes the adulterer. When he sent the drawing to Smithers he explained, 'I am doing the adulterer fiddling with his foreskin in impatient expectation—rather a nice picture.'[21] The adulterer, in a wig similar to the courtier's but dressed in extreme dishabille, wearing only a filmy blouse, bends forward to peep through the bedcurtains in the same impatient posture as the courtier of the frontispiece: they are the same man, though as the adulterer his features are transformed by a coarse and bestial leer. How graphically this reveals the erotic foundation of the over-dressed eighteenth-century figures that Beardsley designed for *The Rape of the Lock*!

Although most of the poem's previous illustrators had costumed

21. *Letters*, p. 155; also pp. 150, 263. Neither the translation nor the drawing was published in his day. In 1976 the drawing was reproduced by Simon Wilson in *Beardsley* (Plate 40).

The Rape of the Lock

AN HEROI-COMICAL POEM

IN FIVE CANTOS

WRITTEN BY

ALEXANDER POPE

EMBROIDERED WITH NINE DRAWINGS

BY

AUBREY BEARDSLEY

"Nolueram, Belinda, tuos violare capillos;
Sed juvat, hoc precibus me tribuisse tuis."—MART.

"A tonso est hoc nomen adepta capillo."—OVID.

LONDON
LEONARD SMITHERS
ARUNDEL STREET
MDCCCXCVI

44. Title-page of *The Rape of the Lock* (1896)

45. Aubrey Beardsley, 'The Dream', frontispiece to *The Rape of the Lock* (1896), photoengraving

46. Aubrey Beardsley, 'The Impatient Adulterer' (1896), drawing

the characters and furnished the settings in the approximate style of Queen Anne's period, Beardsley contrived an idiom that was eclectic, mainly eighteenth-century French rococo spiced with bits of symbolist decoration and art nouveau. The men's clothes—this courtier's, for example—are modeled on ballet dancers' costumes of the first half of the eighteenth century, the skirts enriched with garlands of roses popular in French court dress of the 1770s. (The women's clothes throughout are in the mode of the 1780s.)[22]

The second drawing, entitled 'The Billet-Doux', is a headpiece that introduces the first canto of the poem. [*Plate 47*] Belinda, in severely simplified outline, is seen against an elaborately decorated art-nouveau bed and wall. Her head-dress as well as the pillows suggests a butterfly, and the headboard of the bed a fan—appropriately symbolic. At first glance the drawing has the innocence of a children's book illustration until one notices that Belinda's breast is naughtily exposed.

In the next drawing, entitled 'The Toilet', Belinda sits at her dressing table. [*Plate 48*] The depiction of such scenes, a frequent subject for Beardsley, has been called 'an indication of narcissistic and transvestite moods'.[23] The text of Pope's poem, where that scene is elaborately developed, presented Beardsley with a suitable text.

In the profusion of what Pope calls 'Unnumber'd Treasures' the pair of scissors visible in the right corner hints of the rape-to-come. Above the scissors, and parallel to the candlestick, stands what at first seems to be a bottle but is actually a tiny male figure wearing a turban; and the rows of buttons on his chest indicate the dress of a page-boy. This is the first unequivocal appearance of an important motif to be repeated in later drawings—a dwarf page-boy. Here he is miniaturized among the other instruments of Belinda's proud vanity. Beardsley's preference for the 'artificial' rather than the 'natural' can be seen in the background, where what at first appears to be the garden visible through the window is really a decorated screen, a *trompe-l'œil* that sustains the hot-house atmosphere of the setting. The fake garden, with its cupola-topped pavilion, is sketched with light stippling; it derived from his memory rather than imagination. A few years earlier (in 1893) he had gone to Paris with Joseph Pennell, the illustrator and an early champion of his work. 'He came over', Pennell later recalled, 'to study French gardens, the

22. Brian Reade, *Aubrey Beardsley* (New York, 1967), p. 352.

23. Ibid., p. 18. See also Malcolm Easton, *Aubrey and the Dying Lady: A Beardsley Riddle* (1972), pp. 122–5.

47. Aubrey Beardsley, 'The Billet Doux', *The Rape of the Lock* (1896), photoengraving

Luxembourg, Versailles and the palaces of St. Cloud and St. Germain. These he got not on paper, but in his head', and used them later in *The Rape of the Lock*.[24]

The next drawing 'The Baron's Prayer' [*Plate 49*] is closely based on the text of the poem. Since he is imploring heaven 'ere *Phœbus* rose', the Baron appropriately wears a dressing-gown and night-cap as he kneels before his altar erected to love. The pyramid is constructed of 'twelve vast *French* Romances, neatly gilt', although a bibliographer might quibble that such romances were usually multi-volumed. They are topped by trophies of the Baron's former conquests, love-fetishes like gloves, ribbons, garters. The clouds of flame that rise above the altar are decorative and symbolic rather than true fire-flames, for only the tall candle standing next to it is actually burning, unlike the multitude of unlit candles elsewhere in the drawings. Nature, as seen in the background, is again artificial, a tapestry this time; and on it is stippled an informal country landscape, with a turreted house visible beyond the trees. The scene in the tapestry points forward to the country setting of Hampton Court, where the Baron will next make his appearance.

24. Joseph Pennell, *Aubrey Beardsley and Other Men of the Nineties*, Pennell Club Book 3 (Philadelphia, 1924), p. 25.

48. Aubrey Beardsley, 'The Toilet', *The Rape of the Lock* (1896), photoengraving

49. Aubrey Beardsley, 'The Baron's Prayer', *The Rape of the Lock* (1896), photoengraving

In 'The Barge' [*Plate 50*] Belinda, the glittering cross visible on her breast, holds court on the elaborately constructed and decorated poop-deck of a boat. The setting is more like a theatre-box, with two beaux in attendance on a proud beauty, its artificial atmosphere sustained by the indistinctness of river and countryside beyond. In calling this drawing 'The Barge', a term Pope does not use in the poem, Beardsley probably wished to invoke the famous one of Shakespeare's Cleopatra; and the Egyptian queen's two 'dimpled boys, like smiling Cupids' have been coalesced into the dwarf-like page-boy (on the far left), who attends Belinda. He has been magnified from the tiny figure on Belinda's dressing-table to become an increasingly significant character in the visual drama, though he is unmentioned in the text of the poem.

The composition as a whole separates into two distinct halves, upper and lower, unified by several common decorative motifs (stippled baskets, tassels, and rose-bud designs). The lower half can be seen, in Freudian terms, as Belinda's libido, for its decorations are unmistakably erotic. From the mouths of the three sun-images hang phallic and testicular forms connected to each other by festoons of rosebuds which themselves are genitalian. The lightly stippled forms are sexual: breast-nipples, apertures vaginal or anal, and phalluses, particularly the fragment of ship's oar (in the lower left), which projects like a distorted penis. Even the baskets of flowers with their pendant tassels, encircled by the same stippled line, give off a lascivious air. Presiding over this riotous display of sexuality sits the heroine, prim and demurely remote, seemingly impervious to it.

In the climactic scene of the poem, 'The Rape of the Lock' [*Plate 51*], Beardsley relegates the main action to the edge of the drawing while he focuses our attention on the winking dwarf page-boy, who is slyly helping himself from the tea-table, beneath which lie some discarded playing cards. In effect this transposition of emphasis becomes a satiric comment. The pair of scissors held by the Baron is the same we saw on Belinda's dressing-table; Beardsley thus brings to light a submerged motif of the poem by implicating Belinda in her own ruin. Wearing elaborate court dress and full-bottomed wig the Baron at first glance looks unlike the Baron at his prayer (in dressing-gown and nightcap); and to be sure, when Beardsley sent this drawing to London, Smithers raised the question. 'Why the (?) after Baron,' Beardsley replied. 'If you compare the faces in the two pictures you will find them surprisingly similar.'[25] The other courtier

25. *Letters*, p. 116.

50. Aubrey Beardsley, 'The Barge', *The Rape of the Lock* (1896), photoengraving

51. Aubrey Beardsley, 'The Rape of the Lock', *The Rape of the Lock* (1896), photoengraving

in the scene, the portly man at the back, who conspicuously sports a cane, must be Sir Plume.

Like an inventive stage designer Beardsley changed the characters' costumes for each scene. Here Belinda, who on the barge wore a light-coloured, gaily embroidered gown, now sits as though propped up by a gown solidly black except for the light relief of random feather-swirls, while its shape as it rests on the floor is that of a gigantic *derrière*. Through the window can be seen a landscape from nature, but nature severely formalized as an *allée* of trees. The urn, just visible to the left, is the kind that stands in front of the Wren palace; Beardsley, in fact, was fond of what he called (in a letter to a friend) 'the ever gracious Hampton Court'.[26]

In 'The Cave of Spleen' Beardsley's genius for the grotesque and the witty has its fullest opportunity. [*Plate 52*] One reviewer, after seeing the published book, singled out this plate as 'a dream of the most fantastic ingenuity and invention, a dream in which Mr. Beardsley, with his fertile capacity for intricate and evil imaginings, revels delightedly'.[27] The goddess Spleen herself reclines in the centre on the right, a faint scowl on her unlined face. Her pendant lock of hair, a small one to be sure, reminds us of Belinda's ravished one. Hair as well as peacock feathers and plumes make up the pervasive textures in this drawing. Spleen's gown suggests heavy, dense hair; several other figures are *dressed* in hair; and even the backdrop, its narrow, vertical lines topped by heavy loops, seems textured like hair. Spleen's cave, in short, is a grotto of hair. The peacock feathers, either on wings (as in the upper centre and the lower right) or detached and looking like baleful eyes (in the upper left), are appropriate symbols in a poem that celebrates the vanity of a London belle. They also testify to Beardsley's fascinated interest in the peacock as a decorative theme. Its popularity had begun with the Pre-Raphaelites, and with added interest from Japanese art had become popular in the Aesthetic Movement during the decades before Beardsley.[28]

Beardsley had read the text of Pope's poem carefully, for he shows Spleen with only two handmaids, which was evidently Pope's

26. *Letters*, p. 153.
27. *Saturday Review*, 83 (1897), 426.
28. Robin Spencer, *The Aesthetic Movement* (1972), pp. 70–1. The most famous surviving example of the peacock-vogue is Whistler's Peacock Room (1876), now in the Freer Gallery, Washington, D.C.

52. Aubrey Beardsley, 'The Cave of Spleen', *The Rape of the Lock* (1896), photoengraving

intention, although illustrators after 1714 had misread him to depict four.[29] At Spleen's right stands Ill-nature, clutching a prayerbook and literally showing 'her Bosom [fill'd] with Lampoons', sheets of rolled paper. Affectation, who stands below her, is a curiously stunted creature, her oversized child's head partially obscured by a veil and set on a mature-breasted body. (This may be Beardsley's interpretation of Pope's lines on Affectation, who 'Shows in her Cheek the Roses of Eighteen,/Practis'd to Lisp, and hang the Head aside'.) Umbriel, the large, full-length figure on the left, has just arrived carrying a branch of spleenwort. With his elaborate turban surmounted by luxuriant plumes and his feminine hips, plump thighs, and spavined legs, he is an androgynous creature who resembles Pope's Umbriel only in his function as the mischievous gnome from the world above. Graphically he resembles Beardsley's own drawing, 'The Abbé', published only a few months earlier in *The Savoy* No. 1.[30]

In this, the most ornate, complex drawing of the set, Beardsley could justify his fancy and fantasy because of Pope's couplet: 'Unnumber'd Throngs, on ev'ry side are seen/Of Bodies chang'd to various Forms by *Spleen*.' In the bottom left corner stands the rotund figure of a man, probably to illustrate the phrase 'Here sighs a Jar'; and a closer inspection reveals that he also embodies the line 'Men prove with Child, as pow'rful Fancy works'. For inside the man's distended abdomen can be seen the form of a foetus, outlined in the same light stipple that Beardsley had used in 'The Barge' for his lascivious flower baskets and tassels. (It is true that the foetus is a recurring motif in Beardsley's drawings throughout his career, but in this instance Pope's line provided him with a text.) A further surprise is that this same figure wears a monocle in his right eye; he is perhaps a caricature of the politician Joseph Chamberlain, whose portraits usually identify him by his single eye-glass.[31]

Next to the pregnant man-jar stand the two 'living *Teapots*' of Pope's text. The one at the left carries in his thigh a foetus, though smaller than the one in the pregnant man-jar. In this, Beardsley borrowed from his own illustration for *Lucian's True History*, 'Birth from the Calf of the Leg', drawn two years earlier but not published.[32] To the right of both tea-pots sits a silhouetted figure

29. See pp. 14, 27 above.
30. Reade, *Beardsley* (1967), plate 423.
31. Brigid Peppin, *Fantasy: Book Illustration 1860–1920* (1975), p. 15.
32. Reade, *Beardsley* (1967), plate 255.

inside a gazebo-like cage: this may be one of the 'Angels in Machines'. But the large female figure in the lower right corner has no model in the text: she is related to both a Sphinx (with peacock wings and without a lion's body) and a Siren (of the type that had a fishtail). She is a zoological freak, resembling the unearthly creatures who inhabited French formal gardens in the seventeenth and symbolist art in the nineteenth century. Below this female figure stands the small, talking goose[berry] pie, and next to it Beardsley's joke for the line: 'Maids turn'd Bottels, call aloud for Corks': a maid seated inside a bottle already corked. (Possibly, too, the large female figure, her body twisted invitingly, illustrates that line.) In the lower right corner and leaning forward stands the dwarf, again in page-boy's costume.

Among some of the other 'Unnumber'd Throngs' in the drawing who have no counterparts in the text of the poem is the small male bust, slightly above the centre of the drawing, head in half profile, and wearing a morning cap. [*Plate 53*] It is Alexander Pope himself, in a pose from the well-known portrait by Godfrey Kneller. With this detail Beardsley paid his wittiest and sincerest homage: the poet himself in the Cave of Spleen.

For his eighth drawing, entitled 'The Battle of the Beaux and Belles' [*Plate 54*], Beardsley reviewed or remembered what Du Guernier had designed for the 1714 edition, for he copied from it the motif of a chair on its side to indicate the battle's violence. Where Du Guernier shows 'fierce *Belinda*' threatening to stab with her deadly bodkin the Baron, who has fallen to the floor, Beardsley shows her, armed with a folded fan, glaring angrily at the Baron, who kneels before her as though begging forgiveness. This confrontation Beardsley imitated from Stothard's 1798 illustration. The whole scene, instead of rocking with violence, is static, fixed as though in a snapshot—or an embroidery frame. The dwarf page-boy, firmly centred in the drawing, attentively watches the encounter between his mistress and the ravisher of her lock.

Since Beardsley invented the dwarf page-boy for this set of drawings and used him so frequently—in five of the seven full-page ones (and perhaps lurking among the decorative motifs in the others)—one wonders what this figure signifies. Most simply, he can be seen as an accessory to eighteenth-century décor, when wealthy and aristocratic ladies sometimes equipped themselves with such servants. But his part in the pictorial drama seems too prominent for that casual function: first as an object on Belinda's dressing-table,

53. Aubrey Beardsley, [portrait of Alexander Pope], detail of plate 52

54. Aubrey Beardsley, 'The Battle of the Beaux and Belles', *The Rape of the Lock* (1896), photoengraving

then (lips pursed) facing her on the barge, then mischievously winking at the reader when the Baron is about to violate Belinda's lock, then (in the Cave of Spleen) bending toward the reader and perhaps guarding the maid inside the corked bottle, and now—finally—looking on with evident satisfaction as Belinda glares at the Baron. He may thus be cast in the priapic role that dwarfs traditionally play in amorous history. Pope, who was only four and a half feet tall, sometimes referred to himself as a dwarf; and once in his epistolary infatuation with Lady Mary Wortley Montagu invoked an erotic image when he wrote to her—she was on her way to Turkey—that if encouraged he would meet her in Lombardy, 'the Scene of those celebrated Amours between the fair Princess and her Dwarf'.[33] In the text of *The Rape of the Lock* Belinda's most prized possession, her dog Shock, may have this priapic function;[34] and perhaps the intrusive dwarf page-boy added by Beardsley is the surrogate for Shock, who never appears in these drawings.[35]

'The New Star', the title Beardsley gave to the final drawing [*Plate 55*], reasserts the conclusion of Pope's poem: that the transitory, ravished lock will add eternal glory to the heavens. Directing the reader's eye to the star is a man in the carnival costume of Louis XIV's court. Beardsley had intended to draw sylphs in the *cul-de-lampe*, as he wrote to Smithers; and in his list of drawings he called the ninth: '*Sylphs* (cul de lampe)'.[36] What his sylphs looked like is an intriguing question, but without an answer since he discarded them. In the lower left of this final drawing he added his initials, the only signature in the series of plates.

He aptly described these nine drawings as 'embroidered'. Except for carefully balanced patches of white every bit of surface is textured like hand embroidery of various sorts, from faintly visible, delicate stippling of dots to heavy, almost solid blacks. In his virtuoso use of black against white Beardsley played so many changes that he made it equivalent to colour. His ingredients were simple enough: dots, lines, and masses of black on white or white on black, with infinite subtleties and gradations. It is also to be marvelled that with only black and white, and without even grey wash, he was able to achieve

33. *Correspondence*, ed. G. Sherburn (Oxford, 1956), I, 365.

34. See page 11 n.1 above.

35. In 1898 Beardsley drew 'The Lady with the Monkey', an illustration for *Mademoiselle de Maupin* (Reade, *Beardsley*, 1967, plate 492). The monkey here can be seen as an amalgam of foetus and dwarf.

36. *Letters*, p. 117.

55. Aubrey Beardsley, 'The New Star', *The Rape of the Lock* (1896), photoengraving

the depth of spatial relationships so that unlike the flatness of Japanese prints—which influenced his earlier style, particularly his *Salomé*—these drawings construct a world that, however unreal in other ways, seems to be three-dimensional.

Granted that Beardsley was a consummate artist in creating this graphic idiom he also had the benefit of the advanced technology of book illustration of his day. Previous to his era the artist or his copyist had to transfer the drawing by hand onto the plate—copper, steel, stone, or wood—from which the illustration would be reproduced; but after about 1865 the artist could design his line drawing on paper, and it could then be photographed directly onto the wood block. This innovation enabled the artist to make his design in any size he wished since it could be reduced by photography to the exact dimensions required for the book, and it could also be reversed (a tedious process by previous methods).[37]

Of Beardsley's ten designs for *The Rape of the Lock* (including the binding) six of the original drawings survive; all of them are larger than their printed counterparts. In print they were reduced by about one fifth, except for the *cul-de-lampe* ('The New Star'), which was reduced by about half. This had two distinct advantages for Beardsley: he could pack into the large drawings more detail than if he had been confined to the book-page size; and the published versions are not approximations in another medium (by himself or a craftsman) but exact copies reduced in size. His own genius and this fortunate technology enabled him to develop the potential of black and white drawing to its utmost limits.

While the book was being printed Beardsley remained in Brussels, and on about 1 May wrote to Smithers about its last drawing: 'I hope you liked the cul de lampe. I thought it rather pretty.'[38] Smithers had apparently been too busy to send him his opinion, and besides, the book was in the press. On May 5th, when Beardsley returned to London—so ill that his mother had gone over to accompany him home—the book was already published, an ordinary edition of 500

37. Joseph Pennell, *The Illustration of Books* (1896), pp. 68–9. Beardsley 'accepted wholeheartedly from the beginning the photo-engraved line-block' (John Russell Taylor, *The Art Nouveau Book in Britain*, 1966, pp. 93–4).

38. *Letters*, p. 128. On 4 May, Beardsley, still in Brussels on the eve of his departure for England, sent Smithers a two-page letter—whose present whereabouts cannot be traced—'relating to his work' and mentioning *The Rape of the Lock* (*An Exhibition of Original Drawings by Aubrey Beardsley with a forword by Joseph Pennell*, May 1 to May 17, 1919, The Rosenbach Galleries, Philadelphia).

copies on paper (with the plates on imitation Japanese vellum) and 25 copies on Japanese vellum.[39] He immediately sent a presentation copy of the *de luxe* edition to Edmund Gosse, along with a letter. 'Mon cher Maître,' he began, 'It was not without hesitation that I allowed myself the pleasure of placing this little edition of *The Rape of the Lock* under your protection,'—it is dedicated to Gosse—'for I feared you would find it a very poor offering. Please accept it as a friend rather than a critic, and forgive if you can some of its shortcomings.'[40] Gosse rose to the occasion with a gracious acceptance. 'My dear Aubrey,' he replied, 'How am I to find words to thank you for so kind a compliment and so exquisite a gift? I am immensely flattered & gratified. I do not think you have ever had a subject which better suited your genius, or one in the "embroidery" of which you expended more fanciful beauty. I am truly proud to be connected with such an ingenious object.'[41]

Beardsley also sent a presentation copy of the book to an admirer of a very different kind. Yvette Guilbert, the celebrated *diseuse*, who was a devoted friend of Arthur Symons, had first played in London in 1894, and she returned in the following four years, staying (appropriately) at the Savoy Hotel.[42] On 6 May she received from Symons the first two issues of *The Savoy*, the second of which contained the specimen drawing from *The Rape of the Lock* (the 'rape' scene). She immediately and impetuously thanked him in a charming *mélange* of English and French. 'I receive just now your Savoy and have immediately *regardé* les images! More I see what does Aubrey Beardsley more I am in love with him! What a talent of conception he has! et quelle élégance, quelle *distinction* dans les lignes de ses dessins. *C'est admirable*! Je voudrais bien le connaitre, et lui dire toute mon admiration. Ou est son *studio*? Encore merci, Cher Monsieur. Je vais demain commencer a lire les 2 books—Je voudrais avoir votre autographe sur ces livres, et je voudrais bien aussi que Aubrey B— me donne un souvenir, un *tout-tout*-petit croquis— pour le mettre avec mes autres Souvenirs d'artiste— Est-ce un homme aimable? On le dit— Bonnes Amities, Yvette Guilbert.'[43]

It is doubtful that Beardsley sent his admirer one of his sketches,

39. Notice by Smithers in 1897 edition.
40. *Letters*, p. 131. Goose's copy of the book is now in the Princeton University Library. Beardsley also dedicated the 1897 'bijou' edition to him.
41. 16 May 1896, MS in Princeton University Library.
42. Yvette Guilbert, *La Passante émerveillée* (Paris, 1929). pp. 156–7, 166–8.
43. MS in the John Rylands University Library of Manchester.

even a tiny one, as she so adroitly requested. But he did ask Smithers (on 8 May) to send a copy of his book, and cautiously instructed him: 'If you think this quite unnecessary and foolish don't trouble to pack and post.'[44] Smithers did send a copy—of the ordinary edition—along with Beardsley's inscribed card, which reads: 'a mademoiselle Yvette Guilbert/hommage de Aubrey Beardsley.' Although the pages of the book are uncut, Mlle Guilbert could still have admired the seven full plates, since they are independently bound into the volume.[45]

The greatest encomium, however, Beardsley heard with his own ears from a leading artist of the time, a great judge of excellence and originality in art. The American-born James McNeill Whistler, who lived in London, kept abreast of everything new in the art world. He disliked *The Yellow Book* as much as its art editor's own work, especially the *Salomé* illustrations. Famous for his acerbic wit, he once scolded his close friend Joseph Pennell for his friendship with Beardsley. 'Why do you get mixed up with such things?' he asked. 'Look at him! He's just like his drawings, he's all hairs and peacock's plumes—hairs on his head, hairs on his fingers ends, hairs in his ears, hairs on his toes. And what shoes he wears—hairs growing out of them!' Early in 1896, several years after this spirited scolding, while Whistler was visiting Pennell, Beardsley dropped by, carrying (as he usually did) a portfolio of his latest work under his arm. As Pennell relates the episode: 'When Beardsley opened the portfolio and began to show us the *Rape of the Lock*, Whistler looked at them first indifferently, then with interest, then with delight. And then he said slowly, "Aubrey, I have made a very great mistake—you are a very great artist." And the boy burst out crying. All Whistler could say, when he could say anything, was "I mean it—I mean it—I mean it."'[46] To be so praised by so eminent and outspoken an artist explains the burst of emotion from the dandified Beardsley.

Aside from such private tribute, the publication of a new illustrated edition of Pope's famous poem was not made much of in newspapers and magazines. *The Times*, which declared itself 'content merely to mention the ingenious illustrations contributed to a . . .

44. *Letters*, p. 132.
45. Her copy is now in the Beinecke Library, Yale University.
46. Pennell, *Life of Whistler*, 5th ed. (1911), pp. 310–11. Although the exact date of this anecdote is difficult to ascertain, it is undoubtedly true, for Pennell later wrote at least two other versions of it: in *Aubrey Beardsley and Other Men of the Nineties* (1924), pp. 32, 41–2, and in *The Adventures of an Illustrator* (1925), pp. 220–1.

famous poem', literally did just that. But the *Athenaeum*, in a brief unsigned review, expressed the animus that had pursued *The Yellow Book* and its 'decadent' contributors. It was very odd, wrote the reviewer, that the 'peculiar and pseudo-Japanese manner' of Beardsley should be applied to *The Rape of the Lock*. Then after mention of the 'hideous binding', the writer bestowed mixed comment on the illustrations for being done with extraordinary care and labour, but with such weak design. 'As it is', he concluded, 'the pains [Beardsley] has taken to go wrong are nothing less than distressing'.[47] As though to balance this disapproval, *The Saturday Review* praised exactly what the *Athenaeum* had condemned, yet at the same time took a similar moral tack: Beardsley's 'art, exquisite, artificial, limited and diseased as at the root it is, contains an amazing amount of power and originality, a sense of beauty in line and a genius for decoration. . . .'[48]

When Smithers later issued a prospectus to stimulate sales of the book he quoted from the unfavourable as well as favourable notices that had already appeared. Both the *Glasgow Herald* and the *Scotsman* praised the drawings as Beardsley's best work, while *Black & White* praised only the printed text of what would have been a 'magnificent edition' had it not been embroidered with the drawings of Beardsley, 'whose niggling absurdities merely excite laughter'.[49] Evidently Smithers believed that any kind of publicity helps sell a book, but it could not have sold too well, for it was being advertised in successive numbers of *The Savoy* until the last (No. 8) in December 1896.

Having published the quarto edition, Smithers—much as a Lintot or a Tonson might do—decided to issue an inexpensive miniature edition; the idea was entirely his.[50] In April 1897 Beardsley, again in Paris, told his sister, 'I am bringing out a third edition of the *Rape of the Lock*. A tiny pocket edition, for which my pictures are being reduced to well-nigh postage stamps. I am making a coloured fore-page for it. 'Twill be a charmette.' (He also mentioned the coloured frontispiece to a friend.)[51] On the same day, he reassured Smithers that he was working on 'the new *Rape* drawing. It shall be pretty.'[52]

47. *The Times*, 13 June 1896, p. 18; *Athenaeum*, 14 Nov. 1896, p. 682.
48. *The Saturday Review*, 83 (1897), 426; also see p. 103 and n. 27 above.
49. Prospectus in John Johnson Collection, Bodleian Library.
50. Vincent O'Sullivan, *Aspects of Wilde* (New York, 1936), p. 105.
51. *Letters*, pp. 297, 306. By third edition Beardsley meant that the paper and vellum issues of 1896 were two separate editions.
52. *Letters*, p. 298.

He was comfortable in his hotel—on the Quai Voltaire, as before—and in excellent spirits. 'Am longing to see proofs of the *Rapelets*', he told Smithers. Since the new drawing was not ready he must have meant the embroideries and text of the 1896 edition in their miniaturized size. But the new drawing gave him trouble. 'I have thought and wrought over the new *Rape* drawing to an extent. The result of my efforts leads me to decide on making a largish picture for great reduction in the manner of the others and equally elaborate.'[53]

At last, about 27 April, Beardsley sent Smithers the lettering for the new cover as well as the new drawing, about which he was somewhat nervous and uncertain. 'Will this drawing reduce and look at all decent?' he asked. 'I will do another at once if it won't. *Implore* them to make a careful block. There is by the way some *Chinese white* on the drawing so let it not be touched with indiarubber.'[54] Evidently, then, something more elaborate than the original illustrations was planned, yet when published the only new design in this 'bijou edition' is its binding. (The 1896 cover design was reduced in size and reproduced in black and white on page 3, and the original nine designs were repeated in greatly reduced dimensions.) Perhaps Smithers, in order to keep the price of the edition low, decided not to add the new drawing, particularly if it were in colour. Still, Beardsley ungrudgingly approved of the edition.[55]

The prospectus that Smithers issued to win orders for the edition was itself of bijou size, and illustrated with 'The Baron's Prayer'. It describes the edition as '1000 copies on art paper and 50 copies on Japanese vellum, in demy 16mo'.[56] (It is $5\frac{1}{2}$ by 4 inches.) The cloth binding is striking for its bright red colour and elegant, economical design. The single motif that is repeated and joined by an undulating stippled line is richly ambiguous: a floral form or a grinning man's head or a nameless erotic configuration. Like the binding of the quarto edition, writes Peter Conrad, this is also a map of the perverted world inside the covers. In the bijou edition 'this miniaturization adapts the book to the fetishistic littleness of the poem, in which the adult responsibilities of existence, love, worship, executed wretches and regal councils, shrink into inconsequence, while trivia correspondingly dilate into threatening immensity.'[57]

53. *Letters*, pp. 305, 307.
54. *Letters*, p. 309.
55. Sims, p. 40 (note 14 above).
56. Copy of prospectus inserted in *The Savoy* No. 3 in the Huntington Library.
57. *Times Literary Supplement*, 25 March 1977, p. 336.

There was no reason to believe that it would be reviewed any more widely than the quarto edition, particularly since a selection of Beardsley's work, *A Book of Fifty Drawings*, had recently been published. Yet *Punch* magazine, which had not noticed the quarto edition, reviewed the bijou one. As a staunch upholder of middle-class prejudice against the 'decadents', including the artist whom Mr. Punch called Aubrey Beardsley-Weirdsley,[58] its ridicule was relatively mild. ' "Dainty" is the word for it', the review begins. 'Nothing short of the epithet "dainty" can be applied to the little pocket-volume containing *The Rape of the Lock*, illustrated in weirdly-fantastic style by Mr. AUBREY BEARDSLEY. . . .' And it concluded with a recommendation that would have warmed the heart of any publisher: 'The book is a dainty curiosity, and there is not a collector of such literary curios who should be without this latest edition of POPE'S "heroi-comical poem," *The Rape of the Lock*, as published by LEONARD SMITHERS of the Royal Arcade, W.'[59] Compared with Beardsley's previously published drawings, those for Pope's poem seemed to be, on the surface at least, less tainted with poisonous decadence; that may have sweetened Mr. Punch's temper.

The relative innocence of the drawings in that respect can be seen in the vituperative article 'Aubrey Beardsley and the Decadents' by Margaret Armour in *The Magazine of Art*. After a general attack on the *fin-de-siècle* decadents, the author sketches Beardsley's career; then, after praising his *Morte d'Arthur*, she continues: 'A certain grossness, which revolts one even in his treatment of inanimate things, gets free reign in his men and women, notably in those of *The Yellow Book* period; of late, in *The Savoy* and *Rape of the Lock*, we have joyfully hailed an improvement.' Then, although she explains what she detests in *The Yellow Book* ('the hectic vice, the slimy nastiness of those faces') she does not explain how Beardsley has improved. But she has a simple solution for England's problem: 'Why not hoist the Decadents altogether off our shoulders and saddle them on to France? She has a nice broad back for such things, and Mr. Beardsley won't be the last straw by many. . . . art like Beardsley's, so excellent in technique and so detestable in spirit, wakes more repugnance than praise—proves us a nation stronger in ethics than in art.'[60] This

58. *Punch*, 16 May 1896, p. 229.
59. *Punch*, 23 Oct. 1897, p. 185.
60. *Magazine of Art*, 20 (1896), 9–12 *passim*. She was a rival illustrator, of a rather debased art nouveau style.

gentle critic cannot be accused of charity, for since she detected a new tendency toward wholesomeness in Beardsley she should have allowed him time to develop it—instead of proposing to ship him off to France.

Other critics proved more discerning in sensing that despite their superficial prettiness the drawings exhibited Beardsley's naughty wit. In a long thoughtful essay in *The Westminster Magazine*, Henry Melancthon Strong called them the culmination of Beardsley's artistic success, and explains why. But when he turns to the exquisite faces in the drawings, he finds that 'beneath all their elegance and refinement there lurks that same expression of vice—to use by far too harsh a word for suggesting their dainty aberrations from the paths of virtue—from which Beardsley so seldom succeeded in escaping'.[61]

Twentieth-century critics, putting aside moralistic bias, evaluated Beardsley's drawings for *The Rape of the Lock* on other grounds. Holbrook Jackson, in his study of the 1890s, calls Beardsley 'a diabolonian incident in British art', and criticizes his illustrations for overpowering their literature: '*The Rape of the Lock* is superseded, not explained, by Beardsley.'[62] But even if this were true, why can the drawings not be regarded as such without setting them up as equal competitor with the text they illustrate? Jackson's objection to the drawings was contradicted about ten years later by a critic who, after calling these (and the *Volpone* drawings) the culmination of Beardsley's genius, writes that the artist had 'found in the sophisticated convention of the eighteenth century, in the civilized, formal, polished art of the age of Pope, a mood and manner in which he could believe'.[63] While one critic (in 1928) merely praises Beardsley's technical virtuosity in these drawings, another calls them (in 1947) his greatest work as book illustrations.[64] A 'unique tour de force, imaginatively and technically', still another critic writes (in 1966), these drawings 'may be regarded as one of Beardsley's most complete achievements'.[65]

But the pendulum has recently swung, and one critic (in 1972)

61. *Westminister Magazine*, 154 (1900), 88, 92.
62. *The Eighteen Nineties* (1913), p. 124.
63. Osbert Burdett, *The Beardsley Period An Essay in Perspective* (1925), p. 120.
64. Haldane Macfall, *Aubrey Beardsley The Man and His Work* (1928), pp. 84–5; Philip James, *English Book Illustration 1800–1900* (1947), pp. 53, 55.
65. John Russell Taylor, *The Art Nouveau Book in Britain* (1966), pp. 99–100.

denigrates the drawings as 'innocuous' in comparison to the savage irony and erotic extravagance of Beardsley's other series.[66] A similar condemnation comes from the magisterial Kenneth Clark (in 1976), who labels the drawings 'pastiches'; and while granting them wit and technical skill, condemns them for being 'illustrations' and not 'visions'.[67] Yet however original an illustrator may be, however visionary, he must be sensitive to the literary text; and it seems mistaken to expect the *Rape of the Lock* drawings to be like those for *Salomé* or for *Lysistrata*.

The test of a good illustration, in the opinion of George du Maurier, the illustrator and novelist, is that it shall haunt the memory when the letter-press is forgotten.[68] But the letter-press of *The Rape of the Lock* is far from forgotten. Pope's poem, after the critical vicissitudes of more than two and a half centuries, is firmly established as a living classic. In Virginia Woolf's recently published diary, when she read the poem on a train journey (in 1915) it seemed to her '"supreme"—almost super-human in its beauty & brilliancy—you really can't believe that such things are written down.'[69] Beardsley's drawings, then, are well matched by the poem: both haunt the memory, neither can be forgotten. Such a master of sensibility and nuance as Henry James generously admitted the collaboration of illustration with text when he used photographs in *The Golden Bowl* (1904). 'One welcomes illustration, in other words, with pride and joy,' he wrote; 'but also with the emphatic view that, might one's "literary jealousy" be duly deferred to, it would quite stand off and on its own feet and thus, as a separate and independent subject of publication, carrying its text in its spirit, just as that text correspondingly carries the plastic possibility, become a still more glorious tribute.'[70] In that sense Pope's couplets, with their intricately varied subtleties, reflect and recharge Beardsley's black and white designs.

The ultimate tribute to Beardsley's success in these drawings is that since his edition—more than eighty years ago—no artist has

66. Malcolm Easton, *Aubrey and the Dying Lady A Beardsley Riddle* (1972), p. 70.
67. *New York Review of Books*, 9 Dec. 1976, p. 48. It is curious that at the beginning of his essay, as testimony of Beardsley's genius Lord Clarke recounts the touching episode of Beardsley's meeting with Whistler (p. 113 above) but fails to mention that Whistler's admiration was aroused by these very drawings.
68. Cited in *British Book Illustration Yesterday and Today*, commentary by M. C. Salaman, ed. G. Holme (1923), p. 2.
69. *Diary, Vol. I 1915–1919*, ed. A. O. Bell (1977), p. 23.
70. 'Preface', *The Golden Bowl* (1904), p. x.

illustrated, 'adorned', or 'embroidered' the text of Pope's poem. Perhaps his drawings are too intimidating; they give us not only a decorative commentary on a great poem, but, in a different medium, a mock-heroical masterpiece in its own right.

APPENDIX I

Illustrations in German, French, and Italian Translations of *The Rape of the Lock*

As its fame and popularity extended beyond Great Britain *The Rape of the Lock* was translated into several Continental languages, sometimes with the addition of illustrations. In describing these I have omitted editions that copied earlier English or foreign illustrations.

A. *Herrn Alexander Popens Lockenraub, ein scherzhaftes Heldengedicht Aus dem Englischen in deutsch Verse übersetzt, von Luisen Adelgunden Victorien Gottschedinn* (Leipsig, 1744)

Both the translator and the artist who designed the five plates were women; the former was wife of the distinguished scholar and poet Johann Christoph Gottsched, and the latter Anna Maria Werner, who signed her plates Vernerin. For the opening canto [*Plate 56*] the artist cleverly depicts Ariel as the glittering youth in Belinda's morning dream: his wings are not visible to Belinda as they are to the reader, who thus knows his real identity. Although Shock is not present we see instead another of Belinda's pets, her parrot 'Poll', who sits uncaged on a perch, mute on this fateful day (we are told in Canto IV). We next see Belinda on her voyage up the Thames [*Plate 57*], in a sail-boat whose mast is surrounded by sylphs, and followed by a boatful of musicians to play the 'melting music' that Pope specifies. In showing the medieval turreted castles along the shore, the artist has designed buildings like those in her native Germany, very different from the modest villas along the Thames. Her interior scene of the card game and tea-party [*Plate 58*] is more realistic and convincing, for the crowded groups are so busy playing and drinking that the Baron, standing quietly behind Belinda, is able to accomplish his deed. Shock, in the lower right, does not protect his mistress because he is too busy in dalliance that itself may be a symbolic action. For the Cave of Spleen [*Plate 59*] the artist copied Du

56. A. M. Werner and J. B. Bernigeroth, Canto I of *Popens Lockenraub* (1744), engraving

57. A. M. Werner and J. B. Bernigeroth, Canto II of *Popens Lockenraub* (1744), engraving

58. A. M. Werner and J. B. Bernigeroth, Canto III of *Popens Lockenraub* (1744), engraving

59. A. M. Werner and J. B. Bernigeroth, Canto IV of *Popens Lockenraub* (1744), engraving

Guernier's 1714 grotto, adding a few monsters of her own in the upper right. And, finally, in the battle for the lock [*Plate 60*] the artist's most original touch is her literal picture of the lock of hair, visible on a sunbeam, as it is drawn heavenward to become a comet. In the poem only the Muse sees that happen, but this illustrator lets us see for ourselves.

B. *Der Lockenraub ein scherzhaftes Heldengedicht von A. Pope frey und metrisch übersezt von G. Merkel* (Leipsig, 1797)

Although the frontispiece to this translation bears no signature of artist or engraver it is worth noticing. [*Plate 61*] Not only has its designer vividly caught the turmoil that followed the Baron's rash deed, but he (or she) shows Belinda passionately threatening to stab the Baron. The drama has an element of *Sturm und Drang* that seems not quite appropriate to a 'jocose epic freely translated in verse'. More in keeping with that mood is the action of Shock, who shares his mistress's anger and shows it by seizing the Baron's foot with his teeth—a realistic, homely addition to the poem's scenario.

C. *Œuvres complettes d'Alexandre Pope, traduite en francois. Nouvele edition* (Paris, 1779)

The set of plates designed by Clément Pierre Marillier, one of the foremost eighteenth-century French book illustrators, is redolent of the voluptuousness, the silky sensuality of French court life so richly expressed in the paintings of Boucher and Fragonard. At Belinda's *toilette* [*Plate 62*] the sylphs are unabashed cupids, seemingly more intent on pandering to Belinda than protecting her. The boat on which she embarks [*Plate 63*] is festooned with garlands of flowers and provided with melting music (a flautist is visible): Belinda could be embarking for the Isle of Cythera rather than Hampton Court. For the scene of the rape [*Plate 64*] the artist, perhaps remembering Ariel's uncertainty whether Belinda shall break Diana's law 'Or some frail *China* Jar receive a Flaw', shows a porcelain vase shattered on the floor (in the lower left). The Cave of Spleen [*Plate 65*] is no shaggy cave of rock but a classical Roman edifice. Spleen is attended by four handmaids (instead of two), a precedent set by Samuel Wale, the second illustrator of that canto. In the battle scene [*Plate 66*] a genuinely violent altercation seems to be taking place; and Umbriel, large-scaled and bat-winged in the previous scene, has been replaced

60. A. M. Werner and J. B. Bernigeroth, Canto V of *Popens Lockenraub* (1744), engraving

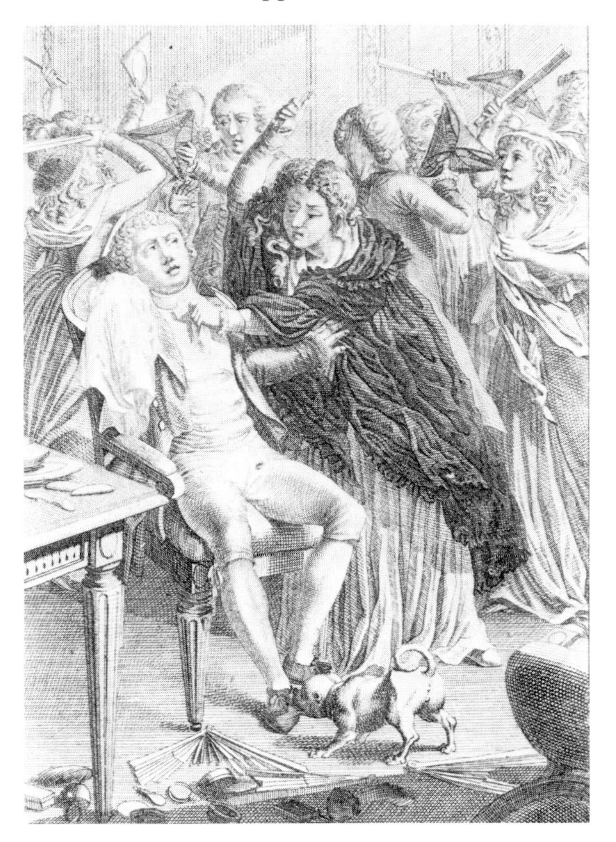

61. Frontispiece to *Der Lockenraub* (1797), engraving

by two gnomes as small as sylphs. Altogether, Marillier has succeeded so well in translating the characters and settings into the French idiom that one is almost startled to remember that he is illustrating an English poem.

D. *Il Riccio Rapito di Alessandro Pope, Tradotto de Antonio Beduschi* (Milano, 1830)

This Italian translation was illustrated with two plates that very economically reduce the drama to middle-class proportions. Belinda is seen adoring the image in her mirror [*Plate 67*], and the Baron celebrates his victory [*Plate 68*] in the corner of a dingy room.

62. C. P. Marillier and P. Duflos *le jeune*, Canto I of 'La Boucle de cheveux enlevée' (1779), engraving

63. C. P. Marillier and Godefroy, Canto II of 'La Boucle de cheveux enlevée' (1779), engraving

64. C. P. Marillier and Dambrun, Canto III of 'La Boucle de cheveux enlevée' (1779), engraving

65. C. P. Marillier and R. de Launoy *le jeune*, Canto IV of 'La Boucle de cheveux enlevée' (1779), engraving

66. C. P. Marillier and L. Halbou, Canto V of 'La Boucle de cheveux enlevée' (1779), engraving

67. De Marchi and Caporali, Canto I of *Il Riccio Rapito* (1830), engraving

68. Pagani and Filippo Caporali, Canto III of *Il Riccio Rapito* (1830), engraving

APPENDIX II

The Rape of the Lock

Text of First Edition (1714)

This is the text for which Du Guernier designed the illustrative plates. It has been generally neglected, and rarely if ever reprinted.

Canto I

What dire Offence from am'rous Causes springs,
What mighty Quarrels rise from trivial Things,
I sing—This Verse to *C---l*, Muse! is due;
This, ev'n *Belinda* may vouchsafe to view:
Slight is the Subject, but not so the Praise,
If She inspire, and He approve my Lays.
Say what strange Motive, Goddess! cou'd compel
A well-bred *Lord* t'assault a gentle *Belle*?
Oh say what stranger Cause, yet unexplor'd,
Cou'd make a gentle *Belle* reject a *Lord*?
And dwell such Rage in softest Bosoms then?
And lodge such daring Souls in Little Men?
Sol thro' white Curtains did his Beams display,
And op'd those Eyes which brighter shine than they;
Now *Shock* had giv'n himself the rowzing Shake,
And Nymphs prepar'd their *Chocolate* to take;
Thrice the wrought Slipper knock'd against the Ground,
And striking Watches the tenth Hour resound.
Belinda still her downy Pillow prest,
Her Guardian *Sylph* prolong'd the balmy Rest.
'Twas he had summon'd to her silent Bed
The Morning Dream that hover'd o'er her Head.
A Youth more glitt'ring than a *Birth-night Beau*,
(That ev'n in Slumber caus'd her Cheek to glow)
Seem'd to her Ear his winning Lips to lay,
And thus in Whispers said, or seem'd to say.
Fairest of Mortals, thou distinguish'd Care

Of thousand bright Inhabitants of Air!
If e'er one Vision touch'd thy infant Thought,
Of all the Nurse and all the Priest have taught,
Of airy Elves by Moonlight Shadows seen,
The silver Token, and the circled Green,
Or Virgins visited by Angel-Pow'rs,
With Golden Crowns and Wreaths of heav'nly Flow'rs,
Hear and believe! thy own Importance know,
Nor bound thy narrow Views to Things below.
Some secret Truths from Learned Pride conceal'd,
To Maids alone and Children are reveal'd:
What tho' no Credit doubting Wits may give?
The Fair and Innocent shall still believe.
Know then, unnumber'd Spirits round thee fly,
The light *Militia* of the lower Sky;
These, tho' unseen, are ever on the Wing,
Hang o'er the *Box*, and hover round the *Ring*.
Think what an Equipage thou hast in Air,
And view with scorn *Two Pages* and a *Chair*.
As now your own, our Beings were of old,
And once inclos'd in Woman's beauteous Mold;
Thence, by a soft Transition, we repair
From earthly Vehicles to these of Air.
Think not, when Woman's transient Breath is fled,
That all her Vanities at once are dead:
Succeeding Vanities she still regards,
And tho' she plays no more, o'erlooks the Cards.
Her Joy in gilded Chariots, when alive,
And Love of *Ombre*, after Death survive.
For when the Fair in all their Pride expire,
To their first Elements the Souls retire:
The Sprights of fiery Termagants in Flame
Mount up, and take a *Salamander*'s Name.
Soft yielding Minds to Water glide away,
And sip with *Nymphs*, their Elemental Tea.
The graver Prude sinks downward to a *Gnome*,
In search of Mischief still on Earth to roam.
The light Coquettes in *Sylphs* aloft repair,
And sport and flutter in the Fields of Air.
  Know farther yet; Whoever fair and chaste
Rejects Mankind, is by some *Sylph* embrac'd:

For Spirits, freed from mortal Laws, with ease
Assume what Sexes and what Shapes they please.
What guards the Purity of melting Maids,
In Courtly Balls, and Midnight Masquerades,
Safe from the treach'rous Friend, and daring Spark,
The Glance by Day, the Whisper in the Dark;
When kind Occasion prompts their warm Desires,
When Musick softens, and when Dancing fires?
'Tis but their *Sylph*, the wise Celestials know,
Tho' *Honour* is the Word with Men below.
   Some Nymphs there are, too conscious of their Face,
For Life predestin'd to the *Gnomes* Embrace.
Who swell their Prospects and exalt their Pride,
When Offers are disdain'd, and Love deny'd.
Then gay Ideas crowd the vacant Brain;
While Peers and Dukes, and all their sweeping Train,
And Garters, Stars, and Coronets appear,
And in soft Sounds, *Your Grace* salutes their Ear.
'Tis these that early taint the Female Soul,
Instruct the Eyes of young *Coquettes* to roll,
Teach Infants Cheeks a bidden Blush to know,
And little Hearts to flutter at a *Beau*.
   Oft when the World imagine Women stray,
The *Sylphs* thro' mystick Mazes guide their Way,
Thro' all the giddy Circle they pursue,
And old Impertinence expel by new.
What tender Maid but must a Victim fall
To one Man's Treat, but for another's Ball?
When *Florio* speaks, what Virgin could withstand,
If gentle *Damon* did not squeeze her Hand?
With varying Vanities, from ev'ry Part,
They shift the moving Toyshop of their Heart;
Where Wigs with Wigs, with Sword-knots Sword-knots strive,
Beaus banish Beaus, and Coaches Coaches drive.
This erring Mortals Levity may call,
Oh blind to Truth! the *Sylphs* contrive it all.
   Of these am I, who thy Protection claim,
A watchful Sprite, and *Ariel* is my Name.
Late, as I rang'd the Crystal Wilds of Air,
In the clear Mirror of thy ruling *Star*
I saw, alas! some dread Event impend,

E're to the Main this Morning's Sun descend.
But Heav'n reveals not what, or how, or where:
Warn'd by thy *Sylph*, oh Pious Maid beware!
This to disclose is all thy Guardian can.
Beware of all, but most beware of Man!
   He said; when *Shock*, who thought she slept too long,
Leapt up, and wak'd his Mistress with his Tongue.
'Twas then *Belinda*! if Report say true,
Thy Eyes first open'd on a *Billet-doux*;
Wounds, *Charms*, and *Ardors*, were no sooner read,
But all the Vision vanish'd from thy Head.
   And now, unveil'd, the *Toilet* stands display'd,
Each Silver Vase in mystic Order laid.
First, rob'd in White, the Nymph intent adores
With Head uncover'd, the *Cosmetic* Pow'rs.
A heav'nly Image in the Glass appears,
To that she bends, to that her Eyes she rears;
Th' inferior Priestess, at her Altar's side,
Trembling, begins the sacred Rites of Pride.
Unnumber'd Treasures ope at once, and here
The various Off'rings of the World appear;
From each she nicely culls with curious Toil,
And decks the Goddess with the glitt'ring Spoil.
This Casket *India*'s glowing Gems unlocks,
And all *Arabia* breaths from yonder Box.
The Tortoise here and Elephant unite,
Transform'd to *Combs*, the speckled and the white.
Here Files of Pins extend their shining Rows,
Puffs, Powders, Patches, Bibles, Billet-doux.
Now awful Beauty puts on all its Arms;
The Fair each moment rises in her Charms,
Repairs her Smiles, awakens ev'ry Grace,
And calls forth all the Wonders of her Face;
Sees by Degrees a purer Blush arise,
And keener Lightnings quicken in her Eyes.
The busy *Sylphs* surround their darling Care;
These set the Head, and those divide the Hair,
Some fold the Sleeve, while others plait the Gown;
And Betty's prais'd for Labours not her own.

Canto II

Not with more Glories, in th'Etherial Plain,
The Sun first rises o'er the purpled Main,
Than issuing forth, the Rival of his Beams
Lanch'd on the Bosom of the Silver *Thames*.
Fair Nymphs, and well-drest Youths around her shone,
But ev'ry Eye was fix'd on her alone.
On her white Breast a sparkling *Cross* she wore,
Which *Jews* might kiss, and Infidels adore.
Her lively Looks a sprightly Mind disclose,
Quick as her Eyes, and as unfix'd as those:
Favours to none, to all she Smiles extends,
Oft she rejects, but never once offends.
Bright as the Sun, her Eyes the Gazers strike,
And, like the Sun, they shine on all alike.
Yet graceful Ease, and Sweetness void of Pride,
Might hide her Faults, if *Belles* had Faults to hide:
If to her share some Female Errors fall,
Look on her Face, and you'll forget 'em all.
This Nymph, to the Destruction of Mankind,
Nourish'd two Locks, which graceful hung behind
In equal Curls, and well conspir'd to deck
With shining Ringlets her smooth Iv'ry Neck.
Love in these Labyrinths his Slaves detains,
And mighty Hearts are held in slender Chains.
With hairy Sprindges we the Birds betray,
Slight Lines of Hair surprize the Finny Prey,
Fair Tresses Man's Imperial Race insnare,
And Beauty draws us with a single Hair.
Th' Adventrous *Baron* the bright Locks admir'd,
He saw, he wish'd, and to the Prize aspir'd:
Resolv'd to win, he meditates the way,
By Force to ravish, or by Fraud betray;
For when Success a Lover's Toil attends,
Few ask, if Fraud or Force attain'd his Ends.
For this, e're *Phœbus* rose, he had implor'd
Propitious Heav'n, and ev'ry Pow'r ador'd,
But chiefly *Love*---to *Love* and Altar built,
Of twelve vast *French* Romances, neatly gilt.
There lay the Sword-knot *Sylvia*'s Hands had sown,

With *Flavia*'s Busk that oft had rapp'd his own:
A Fan, a Garter, half a Pair of Gloves;
And all the Trophies of his former Loves.
With tender *Billet-doux* he lights the Pyre,
And breaths three am'rous Sighs to raise the Fire.
Then prostrate falls, and begs with ardent Eyes
Soon to obtain, and long possess the Prize:
The Pow'rs gave Ear, and granted half his Pray'r,
The rest, the Winds dispers'd in empty Air.
But now secure the painted Vessel glides,
The Sun-beams trembling on the floating Tydes,
While melting Musick steals upon the Sky,
And soften'd Sounds along the Waters die.
Smooth flow the Waves, the Zephyrs gently play[,]
Belinda smil'd, and all the World was gay.
All but the *Sylph*---With careful Thoughts opprest,
Th' impending Woe sate heavy on his Breast.
He summons strait his Denizens of Air;
The lucid Squadrons round the Sails repair:
Soft o'er the Shrouds Aerial Whispers breath,
That seem'd but *Zephyrs* to the Train beneath.
Some to the Sun their Insect-Wings unfold,
Waft on the Breeze, or sink in Clouds of Gold.
Transparent Forms, too fine for mortal Sight,
Their fluid Bodies half dissolv'd in Light.
Loose to the Wind their airy Garments flew,
Thin glitt'ring Textures of the filmy Dew;
Dipt in the richest Tincture of the Skies,
Where Light disports in ever-mingling Dies,
While ev'ry Beam new transient Colours flings,
Colours that change whene'er they wave their Wings.
Amid the Circle, on the gilded Mast,
Superior by the Head, was *Ariel* plac'd;
His Purple Pinions opening to the Sun,
He rais'd his Azure Wand, and thus begun.
Ye *Sylphs* and *Sylphids*, to your Chief give Ear,
Fays, *Fairies*, *Genii*, *Elves*, and *Dæmons* hear!
Ye know the Spheres and various Tasks assign'd.
By Laws Eternal, to th' Aerial Kind.
Some in the Fields of purest *Æther* play,
And bask and whiten in the Blaze of Day.

Some guide the Course of wandring Orbs on high,
Or roll the Planets thro' the boundless Sky.
Some less refin'd, beneath the Moon's pale Light,
Hover, and catch the shooting Stars by Night;
Or suck the Mists in grosser Air below,
Or dip their Pinions in the painted Bow,
Or brew fierce Tempests on the wintry Main,
Or on the Glebe distill the kindly Rain.
Others on Earth o'er human Race preside,
Watch all their Ways, and all their Actions guide:
Of these the Chief the Care of Nations own,
And guard with Arms Divine the *British Throne*.
  Our humbler Province is to tend the Fair,
Not a less pleasing, tho' less glorious Care.
To save the Powder from too rude a Gale,
Nor let th' imprison'd Essences exhale,
To draw fresh Colours from the vernal Flow'rs,
To steal from Rainbows ere they drop in Show'rs
A brighter Wash; to curl their waving Hairs,
Assist their Blushes, and inspire their Airs;
Nay oft, in Dreams, Invention we bestow,
To change a *Flounce*, or add a *Furbelo*.
  This Day, black Omens threat the brightest Fair
That e'er deserv'd a watchful Spirit's Care;
Some dire Disaster, or by Force, or Slight,
But what, or where, the Fates have wrapt in Night.
Whether the Nymph shall break *Diana*'s Law,
Or some frail *China* Jar receive a Flaw,
Or stain her Honour, or her new Brocade,
Forget her Pray'rs, or miss a Masquerade,
Or lose her Heart, or Necklace, at a Ball;
Or whether Heav'n has doom'd that *Shock* must fall.
Haste then ye Spirits! to your Charge repair;
The flutt'ring Fan be *Zephyretta*'s Care;
The Drops to thee, *Brillante*, we consign;
And *Momentilla*, let the Watch be thine;
Do thou, *Crispissa*, tend her fav'rite Lock;
Ariel himself shall be the Guard of *Shock*.
  To Fifty chosen *Sylphs*, of special Note,
We trust th' important Charge, the *Petticoat*:
Oft have we known that sev'nfold Fence to fail,

Tho' stiff with Hoops, and arm'd with Ribs of Whale.
Form a strong Line about the Silver Bound,
And guard the wide Circumference around.
  Whatever Spirit, careless of his Charge,
His Post neglects, or leaves the Fair at large,
Shall feel sharp Vengeance soon o'ertake his Sins,
Be stopt in *Vials*, or transfixt with *Pins*;
Or plung'd in Lakes of bitter *Washes* lie,
Or wedg'd whole Ages in a *Bodkin*'s Eye:
Gums and *Pomatums* shall his Flight restrain,
While clog'd he beats his silken Wings in vain;
Or Alom-*Stypticks* with contracting Power
Shrink his thin Essence like a rivell'd Flower.
Or as *Ixion* fix'd, the Wretch shall feel
The giddy Motion of the whirling Mill,
Midst Fumes of burning Chocolate shall glow,
And tremble at the Sea that froaths below!
  He spoke; the Spirits from the Sails descend;
Some, Orb in Orb, around the Nymph extend,
Some thrid the mazy Ringlets of her Hair,
Some hang upon the Pendants of her Ear;
With beating Hearts the dire Event they wait,
Anxious, and trembling for the Birth of Fate.

Canto III

  Close by those Meads for ever crown'd with Flow'rs,
Where *Thames* with Pride surveys his rising Tow'rs,
There stands a Structure of Majestick Frame,
Which from the neighb'ring *Hampton* takes its Name.
Here *Britain*'s Statesmen oft the Fall foredoom
Of Foreign Tyrants, and of Nymphs at home;
Here Thou, great *Anna*! whom three Realms obey,
Dost sometimes Counsel take---and sometimes *Tea*.
  Hither the Heroes and the Nymphs resort,
To taste awhile the Pleasures of a Court;
In various Talk th' instructive hours they past,
Who gave a *Ball*, or paid the *Visit* last:
One speaks the Glory of the *British Queen*,
And one describes a charming *Indian Screen*;
A third interprets Motions, Looks, and Eyes;

At ev'ry Word a Reputation dies.
Snuff, or the *Fan*, supply each Pause of Chat,
With singing, laughing, ogling, and all that.
  Mean while declining from the Noon of Day,
The Sun obliquely shoots his burning Ray;
The hungry Judges soon the Sentence sign,
And Wretches hang that Jury-men may Dine;
The Merchant from th' *Exchange* returns in Peace,
And the long Labours of the *Toilette* cease ---
Belinda now, whom Thirst of Fame invites,
Burns to encounter two adventrous Knights,
At *Ombre* singly to decide their Doom;
And swells her Breast with Conquests yet to come.
Strait the three Bands prepare in Arms to join,
Each Band the number of the Sacred Nine.
Soon as she spreads her Hand, th' Aerial Guard
Descend, and sit on each important Card:
First *Ariel* perch'd upon a *Matadore*,
Then each, according to the Rank they bore;
For *Sylphs*, yet mindful of their ancient Race,
Are, as when Women, wondrous fond of Place.
  Behold, four *Kings* in Majesty rever'd,
With hoary Whiskers and a forky Beard;
And four fair *Queens* whose hands sustain a Flow'r,
Th' expressive Emblem of their softer Pow'r;
Four *Knaves* in Garbs succinct, a trusty Band,
Caps on their heads, and Halberds in their hand;
And Particolour'd Troops, a shining Train,
Draw forth to Combat on the Velvet Plain.
  The skilful Nymph reviews her Force with Care;
Let Spades be Trumps, she said, and Trumps they were.
Now move to War her Sable *Matadores*,
In Show like Leaders of the swarthy *Moors*.
Spadillio first, unconquerable Lord!
Led off two captive Trumps, and swept the Board.
As many more *Manillio* forc'd to yield,
And march'd a Victor from the verdant Field.
Him *Basto* follow'd but his Fate more hard
Gain'd but one Trump and one *Plebeian* Card.
With his broad Sabre next, a Chief in Years,
The hoary Majesty of *Spades* appears;

Puts forth one manly Leg, to fight reveal'd;
The rest his many-colour'd Robe conceal'd.
The Rebel-*Knave*, that dares his Prince engage,
Proves the just Victim of his Royal Rage.
Ev'n mighty *Pam* that Kings and Queens o'erthrew,
And mow'd down Armies in the Fights of *Lu*,
Sad Chance of War! now, destitute of Aid,
Falls undistinguish'd by the Victor *Spade*!
  Thus far both Armies to *Belinda* yield;
Now to the *Baron* Fate inclines the Field.
His warlike *Amazon* her Host invades,
Th' Imperial Consort of the Crown of *Spades*.
The *Club*'s black Tyrant first her Victim dy'd,
Spite of his haughty Mien, and barb'rous Pride:
What boots the Regal Circle on his Head,
His Giant Limbs in State unwieldy spread?
That long behind he trails his pompous Robe,
And of all Monarchs only grasps the Globe?
  The *Baron* now his *Diamonds* pours apace;
Th' embroider'd *King* who shows but half his Face,
And his refulgent *Queen*, with Pow'rs combin'd,
Of broken Troops an easie Conquest find.
Clubs, *Diamonds*, *Hearts*, in wild disorder seen,
With Throngs promiscuous strow the level Green.
Thus when dispers'd a routed Army runs,
Of *Asia*'s Troops, and *Africk*'s Sable Sons,
With like Confusion different Nations fly,
In various Habits and of various Dye,
The pierc'd Battalions dis-united fall,
In Heaps on Heaps; one Fate o'erwhelms them all.
  The *Knave* of *Diamonds* now exerts his Arts,
And wins (oh shameful Chance!) the *Queen* of *Hearts*.
At this, the Blood the Virgin's Cheek forsook,
A livid Paleness spreads o'er all her Look;
She sees, and trembles at th' approaching Ill,
Just in the Jaws of Ruin, and *Codille*.
And now, (as oft in some distemper'd State)
On one nice *Trick* depends the gen'ral Fate,
An *Ace* of Hearts steps forth: The *King* unseen
Lurk'd in her Hand, and mourn'd his captive *Queen*.
He springs to Vengeance with an eager pace,

And falls like Thunder on the prostrate *Ace*.
The Nymph exulting fills with Shouts the Sky,
The Walls, the Woods, and long Canals reply.
Oh thoughtless Mortals! ever blind to Fate,
Too soon dejected, and too soon elate!
Sudden these Honours shall be snatch'd away,
And curs'd for ever this Victorious Day.
For lo! the Board with Cups and Spoons is crown'd,
The Berries crackle, and the Mill turns round.
On shining Altars of *Japan* they raise
The silver Lamp, and fiery Spirits blaze.
From silver Spouts the grateful Liquors glide,
And *China*'s Earth receives the smoking Tyde.
At once they gratify their Scent and Taste,
While frequent Cups prolong the rich Repast.
Strait hover round the Fair her Airy Band;
Some, as she sip'd, the fuming Liquor fann'd,
Some o'er her Lap their careful Plumes display'd,
Trembling, and conscious of the rich Brocade.
Coffee, (which makes the Politician wise,
And see thro' all things with his half shut Eyes)
Sent up in Vapours to the *Baron*'s Brain
New Stratagems, the radiant Lock to gain.
Ah cease rash Youth! desist e'er 'tis too late,
Fear the just Gods, and think of *Scylla*'s Fate!
Chang'd to a Bird, and sent to flit in Air,
She dearly pays for *Nisus*' injur'd Hair!
But when to Mischief Mortals bend their Mind,
How soon fit Instruments of Ill they find?
Just then, *Clarissa* drew with tempting Grace
A two-edg'd Weapon from her shining Case;
So Ladies in Romance assist their Knight,
Present the Spear, and arm him for the Fight.
He takes the Gift with rev'rence, and extends
The little Engine on his Finger's Ends,
This just behind *Belinda*'s Neck he spread,
As o'er the fragrant Steams she bends her Head:
Swift to the Lock a thousand Sprights repair,
A thousand Wings, by turns, blow back the Hair,
And thrice they twitch'd the Diamond in her Ear,
Thrice she look'd back, and thrice the Foe drew near.

Just in that instant, anxious *Ariel* sought
The close Recesses of the Virgin's Thought;
As on the Nosegay in her Breast reclin'd,
He watch'd th' Ideas rising in her Mind,
Sudden he view'd, in spite of all her Art,
An Earthly Lover lurking at her Heart.
Amaz'd, confus'd, he found his Pow'r expir'd,
Resign'd to Fate, and with a Sigh retir'd.
  The Peer now spreads the glitt'ring *Forfex* wide,
T'inclose the Lock; now joins it, to divide.
Ev'n then, before the fatal Engine clos'd,
A wretched *Sylph* too fondly interpos'd;
Fate urg'd the Sheers, and cut the *Sylph* in twain,
(But Airy Substance soon unites again)
The meeting Points the sacred Hair dissever
From the fair Head, for ever and for ever!
  Then flash'd the living Lightnings from her Eyes,
And Screams of Horror rend th' Affrighted Skies.
Not louder Shrieks by Dames to Heav'n are cast,
When Husbands or when Monkeys breath their last,
Or when rich *China* Vessels, fal'n from high,
In glittring Dust and painted Fragments lie!
  Let Wreaths of Triumph now my Temples twine,
(The Victor cry'd) the glorious Prize is mine!
While Fish in Streams, or Birds delight in Air,
Or in a Coach and Six the *British* Fair,
As long as *Atalantis* shall be read,
Or the small Pillow grace a Lady's Bed,
While *Visits* shall be paid on solemn Days,
When numerous Wax-lights in bright Order blaze,
While Nymphs take Treats, or Assignations give,
So long my Honour, Name, and Praise shall live!
  What Time wou'd spare, from Steel receives its date,
And Monuments, like Men, submit to Fate!
Steel did the Labour of the Gods destroy,
And strike to Dust th' Imperial Tow'rs of *Troy*;
Steel cou'd the Works of mortal Pride confound,
And hew Triumphal Arches to the Ground.
What Wonder then, fair Nymph! thy Hairs shou'd feel
The conqu'ring Force of unresisted Steel?

Canto IV

But anxious Cares the pensive Nymph opprest,
And secret Passions labour'd in her Breast.
Not youthful Kings in Battel seiz'd alive,
Not scornful Virgins who their Charms survive,
Not ardent Lovers robb'd of all their Bliss,
Not ancient Ladies when refus'd a Kiss,
Not Tyrants fierce that unrepenting die,
Not *Cynthia* when her *Manteau*'s pinn'd awry,
E'er felt such Rage, Resentment and Despair,
As Thou, sad Virgin! for thy ravish'd Hair.
For, that sad moment, when the *Sylphs* withdrew,
And *Ariel* weeping from *Belinda* flew,
Umbriel, a dusky melancholy Spright,
As ever sully'd the fair face of Light,
Down to the Central Earth, his proper Scene,
Repairs to search the gloomy Cave of *Spleen*.
Swift on his sooty Pinions flitts the *Gnome*,
And in a Vapour reach'd the dismal Dome.
No cheerful Breeze this sullen Region knows,
The dreaded *East* is all the Wind that blows.
Here, in a Grotto, sheltred close from Air,
And screen'd in Shades from Day's detested Glare,
She sighs for ever on her pensive Bed,
Pain at her side, and *Languor* at her Head.
Two Handmaids wait the Throne: Alike in Place,
But diff'ring far in Figure and in Face.
Here stood *Ill-nature* like an *ancient Maid*,
Her wrinkled Form in *Black* and *White* array'd;
With store of Pray'rs, for Mornings, Nights, and Noons,
Her Hand is fill'd; her Bosom with Lampoons.
There *Affectation* with a sickly Mien
Shows in her Cheek the Roses of Eighteen,
Practis'd to Lisp, and hang the Head aside,
Faints into Airs, and languishes with Pride;
On the rich Quilt sinks with becoming Woe,
Wrapt in a Gown, for Sickness, and for Show.
The Fair ones feel such Maladies as these,
When each new Night-Dress gives a new Disease.
A constant *Vapour* o'er the Palace flies;

Strange Phantoms rising as the Mists arise;
Dreadful, as Hermit's Dreams in haunted Shades,
Or bright as Visions of expiring Maids.
Now glaring Fiends, and Snakes on rolling Spires,
Pale Spectres, gaping Tombs, and Purple Fires:
Now Lakes of liquid Gold, *Elysian* Scenes,
And Crystal Domes, and Angels in Machines.
Unnumber'd Throngs on ev'ry side are seen
Of Bodies chang'd to various Forms by *Spleen.*
Here living *Teapots* stand, one Arm held out,
One bent; the Handle this, and that the Spout:
A Pipkin there like *Homer*'s *Tripod* walks;
Here sighs a Jar, and there a Goose-pye talks;
Men prove with Child, as pow'rful Fancy works,
And Maids turn'd Bottels, call aloud for Corks.
Safe past the *Gnome* thro' this fantastick Band,
A Branch of healing *Spleenwort* in his hand.
Then thus addrest the Pow'r—Hail wayward Queen;
Who rule the Sex to Fifty from Fifteen,
Parent of Vapors and of Female Wit,
Who give th' *Hysteric* or *Poetic* Fit,
On various Tempers act by various ways,
Make some take Physick, others scribble Plays;
Who cause the Proud their Visits to delay,
And send the Godly in a Pett, to pray.
A Nymph there is, that all thy Pow'r disdains,
And thousands more in equal Mirth maintains.
But oh! if e'er thy *Gnome* could spoil a Grace,
Or raise a Pimple on a beauteous Face,
Like Citron-Waters Matron's Cheeks inflame,
Or change Complexions at a losing Game;
If e'er with airy Horns I planted Heads,
Or rumpled Petticoats, or tumbled Beds,
Or caus'd Suspicion when no Soul was rude,
Or discompos'd the Head-dress of a Prude,
Or e'er to costive Lap-Dog gave Disease,
Which not the Tears of brightest Eyes could ease:
Hear me, and touch *Belinda* with Chagrin;
That single Act gives half the World the Spleen.
The Goddess with a discontented Air
Seems to reject him, tho' she grants his Pray'r.

A wondrous Bag with both her Hands she binds,
Like that where once *Ulysses* held the Winds;
There she collects the Force of Female Lungs,
Sighs, Sobs, and Passions, and the War of Tongues.
A Vial next she fills with fainting Fears,
Soft Sorrows, melting Griefs, and flowing Tears.
The *Gnome* rejoicing bears her Gift away,
Spreads his black Wings, and slowly mounts to Day.
  Sunk in *Thalestris*' Arms the Nymph he found,
Her Eyes dejected and her Hair unbound.
Full o'er their Heads the swelling Bag he rent,
And all the Furies issued at the Vent.
Belinda burns with more than mortal Ire,
And fierce *Thalestris* fans the rising Fire.
O wretched Maid! she spread her hands, and cry'd,
(While *Hampton*'s Ecchos, wretched Maid reply'd)
Was it for this you took such constant Care
The *Bodkin*, *Comb*, and *Essence* to prepare;
For this your Locks in Paper-Durance bound,
For this with Tort'ring Irons wreath'd around?
For this with Fillets strain'd your tender Head,
And bravely bore the double Loads of Lead?
Gods! shall the Ravisher display your Hair,
While the Fops envy, and the Ladies stare!
Honour forbid! at whose unrival'd Shrine
Ease, Pleasure, Virtue, All, our Sex resign.
Methinks already I your Tears survey,
Already hear the horrid things they say,
Already see you a degraded Toast,
And all your Honour in a Whisper lost!
How shall I, then, your helpless Fame defend?
'Twill then be Infamy to seem your Friend!
And shall this Prize, th' inestimable Prize,
Expos'd thro' Crystal to the gazing Eyes,
And heighten'd by the Diamond's circling Rays,
On that Rapacious Hand for ever blaze?
Sooner shall Grass in *Hide*-Park *Circus* grow,
And Wits take Lodgings in the Sound of *Bow*;
Sooner let Earth, Air, Sea, to *Chaos* fall,
Men, Monkies, Lap-dogs, Parrots, perish all!
  She said; then raging to *Sir Plume* repairs,

And bids her *Beau* demand the precious Hairs:
(*Sir Plume* of *Amber Snuff-box* justly vain,
And the nice Conduct of a *clouded Cane*)
With earnest Eyes, and round unthinking Face,
He first the Snuff-box open'd, then the Case,
And thus broke out——'My Lord, why, what the Devil?
Z–––ds! damn the Lock! 'fore Gad, you must be civil!
Plague on't! 'tis past a Jest–––nay prithee, Pox!
Give her the Hair'–––he spoke, and rapp'd his Box.
  It grieves me much (reply'd the Peer again)
Who speaks so well shou'd ever speak in vain.
But by this Lock, this sacred Lock I swear.
(Which never more shall joint its parted Hair,
Which never more its Honours shall renew,
Clipt from the lovely Head where once it grew)
That while my Nostrils draw the vital Air,
This Hand, which won it, shall for ever wear.
He spoke, and speaking in proud Triumph spread
The long-contended Honours of her Head.
  But *Umbriel*, hateful *Gnome*! forbears not so;
He breaks the Vial whence the Sorrows flow.
Then see! the *Nymph* in beauteous Grief appears,
Her Eyes half languishing, half drown'd in Tears;
On her heav'd Bosom hung her drooping Head,
Which, with a Sigh, she rais'd; and thus she said.
  For ever curs'd be this detested Day,
Which snatch'd my best, my fav'rite Curl away!
Happy! ah ten times happy, had I been,
If *Hampton-Court* these Eyes had never seen!
Yet am not I the first mistaken Maid,
By Love of *Courts* to num'rous Ills betray'd.
Oh had I rather un-admir'd remain'd
In some lone Isle, or distant *Northern* Land;
Where the gilt *Chariot* never mark'd the way,
Where none learn *Ombre*, none e'er taste *Bohea*!
There kept my Charms conceal'd from mortal Eye,
Like Roses that in Desarts bloom and die.
What mov'd my Mind with youthful Lords to rome?
O had I stay'd, and said my Pray'rs at home!
'Twas this, the Morning *Omens* did foretel;
Thrice from my trembling hand the *Patch-box* fell;

The tott'ring *China* shook without a Wind,
Nay, *Poll* sate mute, and *Shock* was most Unkind!
A *Sylph* too warn'd me of the Threats of Fate,
In mystic Visions, now believ'd too late!
See the poor Remnants of this slighted Hair!
My hands shall rend what ev'n thy own did spare.
This, in two sable Ringlets taught to break,
Once gave new Beauties to the snowie Neck.
The Sister-Lock now sits uncouth, alone,
And in its Fellow's Fate foresees its own;
Uncurl'd it hangs, the fatal Sheers demands;
And tempts once more thy sacrilegious Hands.
Oh hadst thou, Cruel! been content to seize
Hairs less in sight, or any Hairs but these!

Canto V

She said: the pitying Audience melt in Tears,
But *Fate* and *Jove* had stopp'd the *Baron*'s Ears.
In vain *Thalestris* with Reproach assails,
For who can move when fair *Belinda* fails?
Not half so fixt the *Trojan* cou'd remain,
While *Anna* begg'd and *Dido* rag'd in vain.
To Arms, to Arms! the bold *Thalestris* cries,
And swift as Lightning to the Combate flies.
All side in Parties, and begin th' Attack;
Fans clap, Silks russle, and tough Whalebones crack;
Heroes and Heroins Shouts confus'dly rise,
And base, and treble Voices strike the Skies.
No common Weapons in their Hands are found,
Like Gods they fight, nor dread a mortal Wound.
So when bold *Homer* makes the Gods engage,
And heav'nly Breasts with human Passions rage;
'Gainst *Pallas*, *Mars*; *Latona*, *Hermes*, Arms;
And all *Olympus* rings with loud Alarms.
Jove's Thunder roars, Heav'n trembles all around;
Blue *Neptune* storms, the bellowing Deeps resound;
Earth shakes her nodding Tow'rs, the Ground gives way;
And the pale Ghosts start at the Flash of Day!
Triumphant *Umbriel* on a Sconce's Height
Clapt his glad Wings, and sate to view the Fight,

Propt on their Bodkin Spears the Sprights survey
The growing combat, or assist the Fray.
  While thro' the Press enrag'd *Thalestris* flies,
And scatters Deaths around from both her Eyes,
A *Beau* and *Witling* perish'd in the Throng,
One dy'd in *Metaphor*, and one in *Song*.
O cruel Nymph! a living Death I bear,
Cry'd *Dapperwit*, and sunk beside his Chair.
A mournful Glance Sir *Fopling* upwards cast,
Those Eyes are made so killing---was his last:
Thus on *Meander*'s flow'ry Margin lies
Th' expiring Swan, and as he sings he dies.
  As bold Sir *Plume* had drawn *Clarissa* down,
Chloe stept in, and kill'd him with a Frown;
She smil'd to see the doughty Hero slain,
But at her Smile, the Beau reviv'd again.
  Now *Jove* suspends his golden Scales in Air,
Weighs the Mens Wits against the Lady's Hair;
The doubtful Beam long nods from side to side;
At length the Wits mount up, the Hairs subside.
  See fierce *Belinda* on the *Baron* flies,
With more than usual Lightning in her Eyes;
Nor fear'd the Chief th' unequal Fight to try,
Who sought no more than on his Foe to die.
But this bold Lord, with manly Strength indu'd,
She with one Finger and a Thumb subdu'd:
Just where the Breath of Life his Nostrils drew,
A Charge of *Snuff* the wily Virgin threw;
The *Gnomes* direct, to ev'ry Atome just,
The pungent Grains of titillating Dust.
Sudden, with starting Tears each Eye o'erflows,
And the high Dome re-ecchoes to his Nose.
  Now meet thy Fate, th' incens'd Virago cry'd,
And drew a deadly *Bodkin* from her Side.
(The same, his ancient Personage to deck,
Her great great Grandsire wore about his Neck
In three *Seal-Rings*; which after melted down,
Form'd a vast *Buckle* for his Widow's Gown:
Her infant Grandame's *Whistle* next it grew,
The *Bells* she gingled, and the *Whistle* blew;
Then in a *Bodkin* grac'd her Mother's Hairs,

Which long she wore, and now Belinda wears.)
Boast not my Fall (he cry'd) insulting Foe!
Thou by some other shalt be laid as low.
Nor think, to die dejects my lofty Mind;
All that I dread, is leaving you behind!
Rather than so, ah let me still survive,
And burn in *Cupid*'s Flames,---but burn alive.
Restore the Lock! she cries; and all around
Restore the Lock! the vaulted Roofs rebound.
Not fierce *Othello* in so loud a Strain
Roar'd for the Handkerchief that caus'd his Pain.
But see how oft Ambitious Aims are cross'd,
And Chiefs contend 'till all the Prize is lost!
The Lock, obtain'd with Guilt, and kept with Pain,
In ev'ry place is sought, but sought in vain:
With such a Prize no Mortal must be blest,
So Heav'n decrees! with Heav'n who can contest?
Some thought it mounted to the Lunar Sphere,
Since all things lost on Earth, are treasur'd there.
There Heroe's Wits are kept in pondrous Vases,
And Beau's in *Snuff-boxes* and *Tweezer-Cases.*
There broken Vows, and Death-bed Alms are found,
And Lovers Hearts with Ends of Riband bound;
The Courtiers Promises, and Sick Man's Pray'rs,
The Smiles of Harlots, and the Tears of Heirs,
Cages for Gnats, and Chains to Yoak a Flea;
Dry'd Butterflies, and Tomes of Casuistry.
But trust the Muse—she saw it upward rise,
Tho' mark'd by none but quick Poetic Eyes:
(So *Rome*'s great Founder to the Heav'ns withdrew,
To *Proculus* alone confess'd in view.)
A sudden Star, it shot thro' liquid Air,
And drew behind a radiant *Trail of Hair.*
Not *Berenice*'s Locks first rose so bright,
The Skies bespangling with dishevel'd Light.
The *Sylphs* behold it kindling as it flies,
And pleas'd pursue its Progress thro' the Skies.
This the *Beau-monde* shall from the *Mall* survey,
And hail with Musick its propitious Ray.
This, the blest Lover shall for *Venus* take,
And send up Vows from *Rosamonda*'s Lake.

This *Partridge* soon shall view in cloudless Skies,
When next he looks thro' *Galilæo*'s Eyes;
And hence th' Egregious Wizard shall foredoom
The Fate of *Louis*, and the Fall of *Rome*.
  Then cease, bright Nymph! to mourn the ravish'd Hair
Which adds new Glory to the shining Sphere!
Not all the Tresses that fair Head can boast
Shall draw such Envy as the Lock you lost.
For, after all the Murders of your Eye,
When, after Millions slain, your self shall die;
When those fair Suns shall sett, as sett they must,
And all those Tresses shall be laid in Dust;
This Lock, the Muse shall consecrate to Fame,
And mid'st the Stars inscribe *Belinda*'s name!

Index

(R *of* L stands for *The Rape of the Lock*.)